# Hands-On Electronics

Packed full of real circuits to build and test, *Hands-On Electronics* is a unique introduction to analog and digital electronics theory and practice. Ideal both as a college textbook and for self-study, the friendly style, clear illustrations and construction details included in the book encourage rapid and effective learning of analog and digital circuit design theory. All the major topics for a typical one-semester course are covered, including *RC* circuits, diodes, transistors, op amps, oscillators, digital logic, counters, D/A converters and more. There are also chapters explaining how to use the equipment needed for the examples (oscilloscope, multimeter and breadboard), together with pinout diagrams for all the key components referred to in the book.

# Hands-On Electronics

## A One-Semester Course for Class Instruction or Self-Study

**Daniel M. Kaplan**

and

**Christopher G. White**

Illinois Institute of Technology

CAMBRIDGE UNIVERSITY PRESS
Cambridge, New York, Melbourne, Madrid, Cape Town, Singapore,
São Paulo, Delhi, Dubai, Tokyo, Mexico City

Cambridge University Press
The Edinburgh Building, Cambridge CB2 8RU, UK

Published in the United States of America by Cambridge University Press, New York

www.cambridge.org
Information on this title: www.cambridge.org/9780521893510

First published 2003
Fourth printing 2007

*A catalogue record for this publication is available from the British Library*

ISBN 978-0-521-81536-9 Hardback
ISBN 978-0-521-89351-0 Paperback

# Contents

## 12  Monostables, counters, multiplexers, and RAM     155

## 13  Digital↔analog conversion     167

# Figures

# Tables

# About the authors

Dr Daniel M. Kaplan received his Ph.D. in Physics in 1979 from the State University of New York at Stony Brook. His thesis experiment discovered the $b$ quark, and he has devoted much of his career to experimentation at the Fermi National Accelerator Laboratory on properties of particles containing heavy quarks. He has taught electronics laboratory courses for non-electrical-engineering majors over a fifteen-year period at Northern Illinois University and at Illinois Institute of Technology, where he is currently Professor of Physics and Director of the Center for Accelerator and Particle Physics. He also serves as Principal Investigator of the Illinois Consortium for Accelerator Research. He has been interested in electronics since high school, during the junior year of which he designed a computer based on DTL integrated circuits. Over more than twenty-five years in experimental particle physics he has often been responsible for much of his experiments' custom-built electronic equipment. He is the author or co-author of over 150 scientific papers and one encyclopedia article, and co-editor of three books on heavy-quark physics and related fields.

Dr Christopher G. White is Assistant Professor of Physics at Illinois Institute of Technology. He received his Ph.D. in Physics from the University of Minnesota in 1990. He has authored or co-authored over 100 scientific articles in the field of high-energy particle physics, and his current research interests involve neutrinos and hyperons. Dr White is an enthusiastic and dedicated teacher who enjoys helping students to overcome their fear of electronics and to gain both confidence and competence.

# To the Reader

Some of you may be encountering electronic circuits and instruments for the first time. Others may have 'played around' with such stuff if, for example, you were ever bitten by the 'ham radio' bug. In either case, this sequence of laboratory experiments has been designed to introduce you to the fundamentals of modern analog and digital electronics.

We use electronic equipment all the time in our work and recreation. Scientists and engineers need to know a bit of electronics, for example to modify or repair some piece of equipment, or to interface two pieces of equipment that may not have been designed for that purpose. To that end, our goal is that by the end of the book, you will be able to design and build any little analog or digital circuit you may find useful, or at least understand it well enough to have an intelligent conversation about the problem with an electrical engineer. A basic knowledge of electronics will also help you to understand and appreciate the quirks and limitations of instruments you will be using in research, testing, development, or process-control settings.

We expect few of you to have much familiarity with such physical theories as electromagnetism or quantum mechanics, so the thrust of this course will be from phenomena and instruments toward theory, not the other way round. If your curiosity is aroused concerning theoretical explanations, so much the better, but unfamiliarity with physical theory should not prevent you from building or using electronic circuits and instruments.

# Acknowledgments

We are grateful to Profs Carlo Segre and Tim Morrison for their contributions and assistance, and especially to the IIT students without whom this book would never have been possible. Finally, we thank our wives and children for their support and patience. It is to them that we dedicate this book.

# Introduction

This book started life as the laboratory manual for the course Physics 300, 'Instrumentation Laboratory', offered every semester at Illinois Institute of Technology to a mix consisting mostly of physics, mechanical engineering, and aeronautical engineering majors. Each experiment can be completed in about four hours (with one or two additional hours of preparation).

This book differs from existing books of its type in that it is faster paced and goes into a bit less depth, in order to accommodate the needs of a one-semester course covering the elements of both analog and digital electronics. In curricula that normally include one year of laboratory instruction in electronics, it may be suitable for the first part of a two-semester sequence, with the second part devoted to computers and computer interfacing – this scheme has the virtue of separating the text for the more rapidly changing computer material from the more stable analog and digital parts.

The book is also suitable for self-study by a person who has access to the necessary equipment and wants a hands-on introduction to the subject. We feel strongly, and experience at IIT has borne out, that to someone who will be working with electronic instrumentation, a hands-on education in the techniques of electronics is much more valuable than a blackboard-and-lecture approach. Certainly it is a better learning process than simply reading a book and working through problems.

The appendices suggest sources for equipment and supplies, provide tables of abbreviations and symbols, and list recommendations for further reading, which includes chapter-by-chapter correspondences to some popular electronics texts written at similar or somewhat deeper levels to ours: the two slim volumes by Dennis Barnaal, *Analog Electronics for Scientific Application* and *Digital Electronics for Scientific Application* (reissued by Waveland Press, 1989); Horowitz and Hill's comprehensive *The Art of Electronics* (Cambridge University Press, 1989); Diefenderfer and Holton's *Principles of Electronic Instrumentation* (Saunders, 1994);

and Simpson's *Introductory Electronics for Scientists and Engineers* (2nd edition, Prentice-Hall, 1987). There is also a glossary of terms and pinout diagrams for transistors and ICs used within. The reader is presumed to be familiar with the rudiments of differential and integral calculus, as well as with elementary college physics (including electricity, magnetism, and direct- and alternating-current circuits, although these topics are reviewed in the text).

The order we have chosen for our subject matter begins with the basics – resistors, Ohm's law, simple AC circuits – then proceeds towards greater complexity by introducing nonlinear devices (diodes), then active devices (bipolar and field-effect transistors). We have chosen to discuss transistors before devices made from them (operational amplifiers, comparators, digital circuitry) so that the student can understand not only how things work but also why.

There are other texts that put integrated circuits, with their greater ease of use, before discrete devices; or digital circuits, with their simpler rules, before the complexities of analog devices. We have tried these approaches on occasion in our teaching and found them wanting. Only by considering first the discrete devices from which integrated circuits are made can the student understand and appreciate the remarkable properties that make ICs so versatile and powerful. A course based on this book thus builds to a pinnacle of intellectual challenge towards the middle, with the three transistor chapters. After the hard uphill slog, it's smooth sailing from there (hold onto your seatbelts!).

The book includes step-by-step instructions and explanations for the following experiments:

1. Multimeter, breadboard, and oscilloscope;
2. *RC* circuits;
3. Diodes and power supplies;
4. Transistors I;
5. Transistors II: FETs;
6. Transistors III: differential amplifier;
7. Introduction to operational amplifiers;
8. More op-amp applications;
9. Comparators and oscillators;
10. Combinational logic;
11. Flip-flops: saving a logic state;

12. Monostables, counters, multiplexers, and RAM;

13. Digital↔analog conversion.

These thirteen experiments fit comfortably within a sixteen-week semester. If you or your instructor prefers, one or two experiments may easily be omitted to leave a couple of weeks at the semester's end for independent student projects. To this end, Chapter 6, 'Transistors III', has been designed so that no subsequent experiment depends on it; obviously this is also the case for Chapter 13, 'Digital↔analog conversion', which has no subsequent experiment.

As you work through the exercises, you will find focus questions and detailed instructions indicated by the symbol '▷'. Key concepts for each exercise will be denoted by the symbol '•'. Finally, the standard system of units for electronics is the MKS system. Although you may occasionally run across other unit systems, we adhere strictly to the MKS standard.

# 1 Equipment familiarization: multimeter, breadboard, and oscilloscope

In this chapter you will become acquainted with the 'workhorses' of electronics testing and prototyping: multimeters, breadboards, and oscilloscopes. You will find these to be indispensable aids both in learning about and in doing electronics.

### Apparatus required

One dual-trace oscilloscope, one powered breadboard, one digital multimeter, two 10X attenuating scope probes, red and black banana leads, two alligator clips.

## 1.1 Multimeter

You are probably already familiar with multimeters. They allow measurement of voltage, current, and resistance. Just as with wristwatches and clocks, in recent years digital meters (commonly abbreviated to DMM for digital multimeter or DVM for digital voltmeter) have superseded the analog meters that were used for the first century and a half or so of electrical work. The multimeters we use have various input jacks that accept 'banana' plugs, and you can connect the meter to the circuit under test using two banana-plug leads. The input jacks are described in Table 1.1. Depending on how you configure the meter and its leads, it displays
- the voltage difference *between* the two leads,
- the current flowing *through* the meter from one lead to the other, or
- the resistance connected *between* the leads.

Multimeters usually have a selector knob that allows you to select what is to be measured and to set the full-scale range of the display to handle inputs of various size. Note: to obtain the highest measurement precision, set the knob to the lowest setting for which the input does not cause overflow.

**Table 1.1.** Digital multimeter inputs.

| Input jack | Purpose | Limits[a] |
|---|---|---|
| COM | reference point used for all measurements | |
| VΩ | input for voltage or resistance measurements | 1000 V DC/750 V AC |
| mA | input for current measurements (low scale) | 200 mA |
| 10 A | input for current measurements (high scale) | 10 A |

[a]For the BK Model 2703B multimeters used in the authors' labs.

To avoid damaging the meter, be sure to read the safety warnings in its data sheet or instruction booklet.

## 1.2 Breadboard

'Breadboard' may seem a peculiar term! Its origins go back to the days when electronics hobbyists built their circuits on wooden boards. The breadboards we use represent a great step forward in convenience, since they include not only sockets for plugging in components and connecting them together, but also power supplies, a function generator, switches, logic displays, etc.

The exercises that follow were designed using the Global Specialties PB-503 Protoboard. If you do not have access to a PB-503, any suitable breadboard will do, provided you have a function generator and two variable power supplies. Additional components that you will need along the way (that are built into the PB-503) include a 1 k and a 10 k potentiometer, a small 8 Ω speaker, two debounced push-button switches, several LED logic indicators, and several on–off switches.

Fig. 1.1 displays many of the basic features of the PB-503. (For simplicity, some PB-503 features that will be used in experiments in later chapters have been omitted.) While the following description is specific to the PB-503, many other breadboards share some, if not all, of these features. The description will thus be of some use for users of other breadboard models as well.

The breadboard's sockets contain spring contacts: if a bare wire is pushed into a socket, the contacts press against it, making an electrical connection. The PB-503's sockets are designed for a maximum wire thickness of 22 AWG ('American Wire Gauge') – anything thicker (i.e., with *smaller*

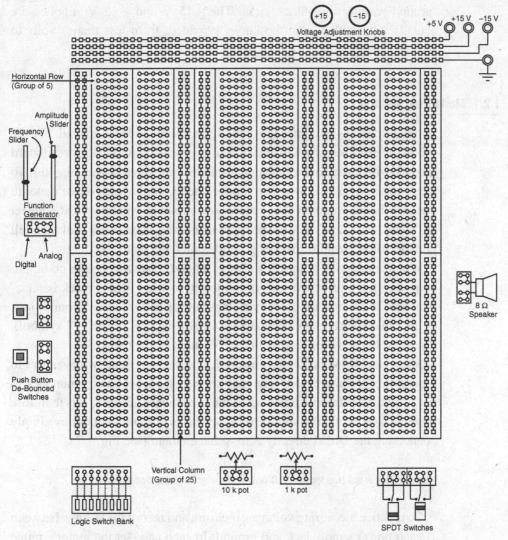

**Fig. 1.1.** Illustration showing many of the basic features of the PB-503 powered Protoboard, with internal connections shown for clarity. Note that each vertical column is broken into halves with no built-in connection between the top and bottom.

AWG number) may damage the socket so that it no longer works reliably for thin wires. The PB-503 sockets are internally connected in groups of five (horizontal rows) or twenty five (vertical columns; see Fig. 1.1).

Each power supply connects to a 'banana' jack and also to a row of sockets running along the top edge of the unit. The three supplies, +5 V (red jack), +15 V (yellow jack), and −15 V (blue jack), have a common 'ground' connection (black jack). The $+15$ V and $-15$ V supplies are actually adjustable, using the knobs provided, from less than 5 volts to greater than 15 volts.

## 1.2.1 Measuring voltage

Voltage is always referenced to something, usually a local ground. For the following exercises you will measure voltage with respect to the breadboard ground, which is also the common ground for the three power supplies. To measure a voltage, you will first connect the 'common' jack of the meter to the breadboard common (i.e., breadboard ground). Next you will connect the meter's 'voltage' jack to the point of interest. The meter will then tell you the voltage with respect to ground at this one point.

When connecting things, it's always a good idea to use color coding to help keep track of which lead is connected to what. Use a black banana-plug lead to connect the 'common' input of the meter to the 'ground' jack of the breadboard (black banana jack labeled with a '$\stackrel{\perp}{\triangledown}$' or '$\stackrel{\perp}{=}$' symbol). Use a red banana-plug lead with the 'V' input of the meter.

Since the DMM is battery powered, it is said to 'float' with respect to ground (i.e., within reason,[1] one may connect the DMM's common jack to any arbitrary voltage with respect to the breadboard ground). It is therefore possible to measure the voltage drop across any circuit element by simply connecting the DMM directly across that element (see Fig. 1.2).

> *Warning:* This is not true for most AC-powered meters and oscilloscopes.

▷ To practice measuring voltages, measure and record the voltage between each power supply jack and ground. In each case set the meter's range for the highest precision (i.e., one setting above overflow).

▷ Adjust the $+15$ V and $-15$ V supplies over their full range and record the minimum and maximum voltage for each. Carefully set the $+15$ V supply to a voltage half-way between its minimum and maximum for use in the next part.

---

[1] If you wonder what we mean by 'within reason', ask yourself what bad thing would happen if you connected the DMM common to, say, twenty million volts – if you're interested, see e.g. H. C. Ohanian, *Physics*, 2nd edition, vol. 2, 'Interlude VI' (Norton, New York, 1988), esp. pp. VI–8 for more information on this.

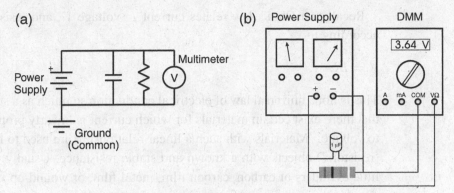

**Fig. 1.2.** Measuring voltage. (a) An arbitrary circuit diagram is shown as an illustration of how to use a voltmeter. Note that the meter measures the voltage drop across both the resistor and capacitor (which have identical voltage drops since they are connected in parallel). (b) A drawing of the same circuit showing how the leads for a DMM should be connected when measuring voltage. Notice how the meter is connected in parallel with the resistor.

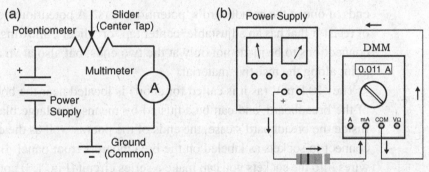

**Fig. 1.3.** Measuring current. (a) Schematic diagram of series circuit consisting of power supply, 10 k potentiometer, and multimeter. (Note that the center tap of the potentiometer is left unconnected in this exercise – accidentally connecting it to power or ground could lead to excessive current flow and burn out the pot.) (b) A drawing of the same circuit showing how the DMM leads should be configured to measure current. Note that the meter is connected in series with the resistor.

## 1.2.2 Measuring current; resistance and Ohm's law

Current is measured by connecting a current meter (an ammeter, or a DMM in its 'current' mode) in series with the circuit element through which the current flows (see Fig. 1.3). Note carefully the differences between Fig. 1.2 and Fig. 1.3.

Recall that Ohm's law relates current $I$, voltage $V$, and resistance $R$ according to

$$V = IR. \tag{1.1}$$

This is not a universal law of electrical conduction so much as a statement that there exist certain materials for which current is linearly proportional to voltage.[2] Materials with such a linear relationship are used to fabricate 'resistors': objects with a known and stable resistance. Usually they are little cylinders of carbon, carbon film, metal film, or wound-up wire, encased in an insulating coating, with wire leads sticking out the ends. Often the resistance is indicated by means of colored stripes according to the resistor color code (Table 1.2). Resistors come in various sizes according to their power rating. The common sizes are $\frac{1}{8}$ W, $\frac{1}{4}$ W, $\frac{1}{2}$ W, 1 W, and 2 W.

You can easily verify this linear relationship between voltage and current using the fixed 10 kΩ (10 000 ohm) resistance provided between the two ends of one of the breadboard's 'potentiometers'. A potentiometer is a type of resistor that has an adjustable 'center tap' or 'slider', allowing electrical connections to be made not only at the two ends, but also at an adjustable point along the resistive material.

The '10 k pot' (as it is called for short) is located near the bottom edge of the breadboard, and can be adjusted by means of a large black knob.[3] Inside the breadboard's case, the ends of the pot (as well as the center tap) connect to sockets as labeled on the breadboard's front panel. By pushing wires into the sockets you can make a series circuit (Fig. 1.3) consisting of an adjustable power supply, the 10 k pot, and the multimeter (configured to measure current). You can attach alligator clips to the meter leads to connect them to the wires. **But**, before doing so, be sure to observe the following warnings:

- First, turn off the breadboard power to avoid burning anything out if you happen to make a mistake in hooking up the circuit.
- Be careful to keep any exposed bits of metal from touching each other and making a 'short circuit'! Note that most of the exposed metal on

---

[2] Of course, the existence of other materials (namely semiconductors) for which the $I$–$V$ relationship is *non*linear makes electronics much more interesting and underlies the transformation of daily life brought about by electronics during the twentieth century.

[3] If you don't have a PB-503 breadboard, find a 10 k pot on your breadboard if it has one; otherwise you will have to purchase a separate 10 k pot.

**Table 1.2.** Color code for nonprecision resistors (5, 10, or 20% tolerance).

The resistance in ohms is the sum of the values in columns 1 and 2, multiplied by the value in column 3, plus or minus the tolerance in column 4. For example, the color code for a 1 k resistor would be 'brown–black–red', for 51 $\Omega$ 'green–brown–black', for 330 $\Omega$ 'orange–orange–brown', etc.

| Stripe: | 1 | 2 | 3 | 4 (tolerance) |
|---|---|---|---|---|
| Black | 0 | 0 | $10^0$ | |
| Brown | 10 | 1 | $10^1$ | |
| Red | 20 | 2 | $10^2$ | |
| Orange | 30 | 3 | $10^3$ | |
| Yellow | 40 | 4 | $10^4$ | |
| Green | 50 | 5 | $10^5$ | |
| Blue | 60 | 6 | $10^6$ | |
| Violet | 70 | 7 | $10^7$ | |
| Gray | 80 | 8 | $10^8$ | |
| White | 90 | 9 | $10^9$ | |
| | | | | |
| Gold | | | | 5% |
| Silver | | | | 10% |
| None | | | | 20% |

Stripe 2

Stripe 1　　Stripe 3

Tolerance Stripe

the breadboard (screw heads for example) has a low-resistance path to ground.

• If you accidentally connect power or ground to the potentiometer's center tap, you can easily burn out the pot, rendering it useless! If in doubt, have someone check your circuit before turning on the power.

▷ Use Ohm's law to predict the current that will flow around the circuit if you use the power supply that you set to its midpoint in the previous exercise. What current should flow if the supply is set to its minimum voltage? What is the current if the supply is set to its maximum voltage?

▷ Now turn on the breadboard power, measure the currents for these three voltages, and compare with your predictions. Make a graph of voltage *vs.* current from these measurements. Is the relationship linear? How close is the slope of voltage vs. current to 10 kΩ?

### 1.2.3 Measuring resistance

Now turn off the breadboard power and disconnect your series circuit. In this and the following part, the pot should connect only to the meter.

▷ Set the meter for resistance and measure and record the resistance between the two ends of your 10 k pot. Due to manufacturing tolerances, you will probably find that it is not exactly 10 kΩ. By what percentage does it differ from the nominal 10 kΩ value? Does the measured value agree more closely with the slope you previously measured than with the nominal value? Explain.

▷ Now connect the meter between the center tap and one end of the pot. What resistance do you observe? What happens to the resistance as you turn the potentiometer's knob?

▷ Leaving the knob in one place, measure and record the resistance between the center tap and each end. Do the two measurements add up to the total you measured above? They should – explain why.

## 1.3 Oscilloscope

With its many switches and knobs, a modern oscilloscope can easily intimidate the faint of heart, yet the scope is an essential tool for electronics troubleshooting and you must become familiar with it. Accordingly, the rest of this laboratory session will be devoted to becoming acquainted with such an instrument and seeing some of the things it can do.

The oscilloscope we use is the Tektronix TDS210 (illustrated in Fig. 1.4). If you don't have a TDS210, any dual-trace oscilloscope, analog or digital, can be used for these labs as long as the bandwidth is high enough – ideally, 30 MHz or higher. While the description below may not correspond exactly to your scope, with careful study of its manual you should be able to figure out how to use your scope to carry out these exercises.

The TDS210 is not entirely as it appears. In the past you may have used an oscilloscope that displayed voltage as a function of time on a

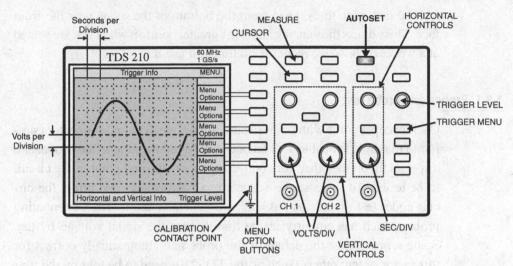

**Fig. 1.4.** Illustration of the Tektronix TDS210 digital oscilloscope. The basic features to be used in this tutorial are marked. Note and remember the location of the 'AUTOSET' button – when all else fails, try autoset!

cathode-ray tube (CRT). While the TDS210 can perform a similar function, it does not contain a CRT (part of the reason it is so light and compact).

Until the 1990s, most oscilloscopes were purely 'analog' devices: an input voltage passed through an amplifier and was applied to the deflection plates of a CRT to control the position of the electron beam. The position of the beam was thus a direct analog of the input voltage. In the past few years, analog scopes have been largely superseded by digital devices such as the TDS210 (although low-end analog scopes are still in common use for TV repair, etc.).

A digital scope operates on the same principle as a digital music recorder. In a digital scope, the input signal is sampled, digitized, and stored in memory. The digitized signal can then be displayed on a computer screen.

One of your first objectives will be to set up the scope to do some of the things for which you may already have used simpler scopes. After that, you can learn about multiple traces and triggering. In order to have something to look at on the scope, you can use your breadboard's built-in function generator, a device capable of producing square waves, sinusoidal waves, and triangular waves of adjustable amplitude and frequency. But start by using the built-in 'calibrator' signal provided by the scope on a metal contact labeled 'probe comp' (or something similar), often located near the lower right-hand corner of the display screen.

Note that a leg folds down from the bottom of the scope near the front face. This adjusts the viewing angle for greater comfort when you are seated at a workbench, so we recommend that you use it.

### 1.3.1 Probes and probe test

Oscilloscopes come with *probes*: cables that have a coaxial connector (similar to that used for cable TV) on one end, for connecting to the scope, and a special tip on the other, for connecting to any desired point in the circuit to be tested. To increase the scope's input impedance and affect the circuit under test as little as possible, we generally use a '10X' attenuating probe, which has circuitry inside that divides the signal voltage by ten. Some scopes sense the nature of the probe and automatically correct for this factor of ten; others (such as the TDS210) need to be told by the user what attenuation setting is in use.

As mentioned above, your scope should also have a built-in 'calibrator' circuit that puts out a standard square wave you can use to test the probe (see Fig. 1.4). The probe's coaxial connector slips over the 'CH 1' or 'CH 2' input jack and turns clockwise to lock into place. The probe tip has a spring-loaded sheath that slides back, allowing you to grab the calibrator-signal contact with a metal hook or 'grabber'.

An attenuating scope probe can distort a signal. The manufacturer therefore provides a 'compensation adjustment' screw, which needs to be tuned for minimum distortion. The screw is usually located on the assembly that connects the probe to the scope, or, occasionally, on the tip assembly.

▷ Display the calibrator square-wave signal on the scope. If the signal looks distorted (i.e., not square), carefully adjust the probe compensation using a small screwdriver. (If you have trouble achieving a stable display, try 'AUTOSET'.)

▷ Check your other probe. Make sure that both probes work, are properly compensated, and have equal calibrations. Sketch the observed waveform.

(Consult your oscilloscope user manual for more information about carrying out a probe test.)

Note that each probe also has an alligator clip (sometimes referred to as the 'reference lead' or 'ground clip'). This connects to the shield of the coaxial cable. It is useful for reducing noise when looking at high-frequency (time intervals of order nanoseconds) or low-voltage signals. Since it is connected directly to the scope's case, which is grounded via the third prong of the AC power plug, it must never be allowed to touch a point in a circuit other than ground! Otherwise you will create a short circuit by connecting multiple points to ground, which could damage circuit components.

This is no trouble if you are measuring a voltage with respect to ground. But if you want to measure a voltage drop between two points in a circuit, neither of which is at ground, first observe one point (with the probe) and then the other. The difference between the two measurements is the voltage across the element. During this process, the reference lead should remain firmly attached to ground and should not be moved! (Alternatively, you can use two probes and configure the scope to subtract one input from the other.)

*Warning:* A short circuit will occur if the probe's reference lead is connected anywhere other than ground.

### 1.3.2 Display

Your oscilloscope user's manual will explain the information displayed on the scope's screen. Record the various settings: timebase calibration, vertical scale factors, etc.

▷ Explain briefly the various pieces of information displayed around the edges of the screen.

The following exercises will give you practice in understanding the various settings. For each, you should study the description in your oscilloscope user's manual. The description below is specific to the TDS210; if you have a different model, your manual will explain the corresponding settings for your scope.

### 1.3.3 Vertical controls

There is a set of 'vertical' controls for each channel (see Fig. 1.4). These adjust the sensitivity (volts per vertical division on the screen) and offset (the vertical position on the screen that corresponds to zero volts). The 'CH 1' and 'CH 2' menu buttons can be used to turn the display of each channel on or off; they also select which control settings are programmed by the push-buttons just to the right of the screen.

▷ Display a waveform from the calibrator on channel 1. What happens when you adjust the POSITION knob? The VOLTS/DIV knob?

### 1.3.4  Horizontal sweep

To the right of the vertical controls are the horizontal controls (see Fig. 1.4). Normally, the scope displays voltage on the vertical axis and time on the horizontal axis. The SEC/DIV knob sets the sensitivity of the horizontal axis, i.e. the interval of time per horizontal division on the screen. The POSITION knob moves the image horizontally on the screen.

▷ How many periods of the square wave are you displaying on the screen? How many divisions are there per period? What time interval corresponds to a horizontal division? Explain how these observations are consistent with the known period of the calibrator signal.

▷ Adjust the SEC/DIV knob to display a larger number of periods. Now what is the time per division? How many divisions are there per period?

### 1.3.5  Triggering

Triggering is probably the most complicated function performed by the scope. To create a stable image of a repetitive waveform, the scope must 'trigger' its display at a particular voltage, known as the trigger 'threshold'. The display is synchronized whenever the input signal crosses that voltage, so that many images of the signal occurring one after another can be superimposed in the same place on the screen. The LEVEL knob sets the threshold voltage for triggering.

You can select whether triggering occurs when the threshold voltage is crossed from below ('rising-edge' triggering) or from above ('falling-edge' triggering) using the trigger menu (or, for some scope models, using trigger control knobs and switches). You can also select the signal source for the triggering circuitry to be channel 1, channel 2, an external trigger signal, or the 120 V AC power line, and control various other triggering features as well.

Since setting up the trigger can be tricky, the TDS210 provides an automatic setup feature (via the AUTOSET button) which can lock in on almost any repetitive signal presented at the input and adjust the voltage sensitivity and offset, the time sensitivity, and the triggering to produce a stable display.

▷ After getting a stable display of the calibrator signal, adjust the LEVEL knob in each direction until the scope just barely stops triggering. What is the range of trigger level that gives stable triggering on the calibrator signal? How does it compare with the amplitude of the calibrator waveform? Does this make sense? Explain.

Next connect the scope probe to the breadboard's function generator – you can do this by inserting a wire into the appropriate breadboard socket and grabbing the other end of the wire with the scope probe's grabber. The function generator's amplitude and frequency are adjusted by means of sliders and slide switches.

▷ Look at each of the waveforms available from the function generator: square, sine, and triangle. Try out the frequency and voltage controls and explain how they work. Adjust the function generator's frequency to about 1 kHz.

▷ Display both scope channels, with one channel looking at the output of the function generator and the other looking at the scope's calibrator signal. Make sure the vertical sensitivity and offset are adjusted for each channel so that the signal trace is visible.

▷ What do you see on the screen if you trigger on channel 1? On channel 2?

▷ What do you see if neither channel causes triggering (for example, if the trigger threshold is set too high or too low)?

▷ How does this depend on whether you select 'normal' or 'auto' trigger mode? Why? (If you find this confusing, be sure to ask for help, or study the oscilloscope manual more carefully.)

## 1.3.6 Additional features

The TDS210 has many more features than the ones we've described so far. Particularly useful are the digital measurement features. Push the MEASURE button to program these. You can use them to measure the amplitude, period, and frequency of a signal. The scope does not measure amplitude directly. How then can you derive the amplitude from something the scope does measure?

▷ Using the measurement features, determine the amplitude, frequency, and period of a waveform of your choice from the function generator.

You can also use the on-screen cursors to make measurements.

▷ Use the cursors to measure the half-period of the signal you just measured.

▷ Explain how you made these measurements and what your results were.

(A feature that comes in particularly useful on occasion is signal averaging; this is programmed via the ACQUIRE button and allows noise, which tends to be random in time, to be suppressed relative to signal, which is usually periodic.)

# 2    RC circuits

Capacitors are not useful in DC circuits since they contain insulating gaps, which are open circuits for DC. However, for voltages that change with time, a simple series circuit with a capacitor and a resistor can output the time derivative or integral of an input signal, or can filter out low-frequency or high-frequency components of a signal. But before plunging into the world of time-varying voltage and current (i.e., *alternating-current* circuits), we explore the *voltage-divider* idea using direct current, since it gives us a simple way to understand circuits containing more than one component in series. Then we apply it to the analysis of *RC* circuits as filters. Note that the series *RC* circuit can be analyzed in two different ways:

• via the exponential charging/discharging equation, and

• as an AC voltage divider.

Both approaches are valid – in fact, they are mathematically equivalent – but the first is more useful when using capacitors as integrators or differentiators, whereas the second is more useful when analyzing low-pass and high-pass filters. The first is referred to as the *time-domain* approach, since it considers the voltage across the capacitor as a function of time, and the second as the *frequency-domain* approach, since it focuses on the filter attenuation *vs.* frequency.

## Apparatus required

Oscilloscope, digital multimeter, breadboard, 68 Ω and 10 kΩ resistors, 0.01 μF ceramic capacitor.

## 2.1  Review of capacitors

As you may recall from an introductory physics course, a capacitor consists of two parallel conductors separated by an insulating gap. The capacitance,

**Table 2.1.** Some typical dielectric materials used in capacitors.

| Material | Dielectric constant ($\kappa$) |
|----------|-------------------------------|
| Vacuum | 1.0 |
| Air (at STP) | 1.00054 |
| Paper | 3.5 |
| Mica | 5.4 |
| Ceramic | $\approx 100$ |

$C$, is proportional to the area of the conductors, $A$, and inversely proportional to their separation, $s$, multiplied by the *dielectric constant*, $\kappa$, of the insulating material:

$$C = \kappa \epsilon_0 A/s,$$

where, in the MKS system of units, $A$ is in meters squared, $s$ in meters, and $C$ in farads, abbreviated F (1 farad $\equiv$ 1 coulomb per volt). (The constant of proportionality is the so-called *permittivity of free space* and has the value $\epsilon_0 = 8.854 \times 10^{-12}$ F/m).

The farad is an impractically large unit: for a conductor area of 1 cm$^2$ and separation of 1 mm, with dielectric constant of order 1, the capacitance is $\sim$ picofarads. To achieve the substantially larger capacitances (of order microfarads) often found in electronic circuits, manufacturers wind ribbon-shaped capacitors up into small cylinders and use insulators of high dielectric strength, such as ceramics or (in the so-called *electrolytic* capacitors) special dielectric pastes, that chemically form an extremely thin insulating layer when a voltage is applied. Table 2.1 gives dielectric constants for some typical dielectrics used in capacitors.

Capacitors thus come in a variety of types, categorized according to the type of dielectric used, which determines how much capacitance can be squeezed into a small volume. Electrolytic and tantalum capacitors are *polarized*, which means that they have a positive end and a negative end, and the applied voltage should be more positive at the positive end than at the negative end – if you reverse-voltage a polarized capacitor it can burn out, or even explode! Paper, mica, and ceramic capacitors are unpolarized and can be hooked up in either direction. The large dielectric constants of the polarized dielectrics permit high capacitance values – up to millifarads in a several-cubic-centimeter can.

## 2.1.1 Use of capacitors; review of AC circuits

The fundamental rule governing the behavior of capacitors is

$$Q = CV, \tag{2.1}$$

where $Q$ is the charge stored on the capacitor at a given time, $V$ is the voltage across the capacitor at that time, and $C$ is the capacitance. Current can flow into or out of a capacitor, but only to the extent that the charge on the capacitor is changing. In other words, the current into or out of a capacitor is equal to the time derivative of the charge stored on it. You can see the resemblance between Eq. 2.1 and Ohm's law (Eq. 1.1). The key difference is that, for a resistor, it is the *time derivative* of the charge that is proportional to voltage, whereas for a capacitor it is the charge itself.

Capacitors are thus useful only in circuits in which voltages or currents are changing in time, namely AC circuits. We will consider the response of circuits to *periodic* waveforms; these can be characterized by their *frequency*, $f$, and *period*, $T$, which of course are related by

$$T = 1/f,$$

as well as their *angular frequency* $\omega = 2\pi f$. (If $f$ is expressed in hertz $\equiv$ cycles per second, then $T$ is in seconds and $\omega$ is in radians per second.) A periodic waveform is also characterized by its *amplitude*, which, assuming the wave is pure AC (i.e. symmetric with respect to ground), is the maximum voltage that it reaches.

There are an infinite variety of AC waveforms, but to understand how capacitors are used, it is sufficient to focus on two: square waves and sine waves. You have already encountered a square wave in the previous chapter – the scope's calibrator signal. A square wave of amplitude $V_0$ is a signal that oscillates back and forth between two voltage levels, one at $+V_0$ and one at $-V_0$, spending 50% of the time at each level. Note that the peak-to-peak voltage is twice the amplitude:

$$V_{\text{p-p}} = 2A.$$

A sine wave is a particularly important case because, by Fourier decomposition, any periodic waveform can be represented as a sum of sine waves of various amplitudes and frequencies. Most of AC circuit analysis therefore concerns itself with the response of circuits to sine waves.

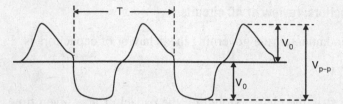

**Fig. 2.1.** Representation of an arbitrary, periodic waveform with period $T$, amplitude $V_0$ and peak-to-peak voltage $V_{p-p}$.

A sine wave can be represented mathematically by

$$V(t) = V_0 \sin(\omega t + \phi). \tag{2.2}$$

If $t$ is in seconds, this describes a voltage with amplitude $V_0$ changing sinusoidally in time at the rate of $\omega$ radians per second. The phase $\phi$ determines the voltage at $t = 0$:

$$V(0) = V_0 \sin \phi. \tag{2.3}$$

Now suppose such a voltage signal is applied to a capacitor; for simplicity, we choose $\phi = 0$, i.e. at $t = 0$ the voltage is zero. Since we are assuming there is no resistance in the circuit, there is no possibility of a voltage drop anywhere in the circuit.[1] Thus, at any moment in time, the voltage across the capacitor must equal the voltage out of the signal source. To find the resulting current we can differentiate Eq. 2.1:

$$I = \frac{dQ}{dt} \tag{2.4}$$

$$= C \frac{dV}{dt} \tag{2.5}$$

$$= C \frac{d}{dt} (V_0 \sin \omega t) \tag{2.6}$$

$$= \omega C V_0 \cos \omega t \tag{2.7}$$

$$= \omega C V_0 \sin (\omega t + 90°). \tag{2.8}$$

We see that the current is also a sine wave, but shifted in time with a phase shift of $90°$, i.e. the current *leads* the voltage by $90°$.

We can write an 'Ohm's-law equivalent' for a capacitor, as long as it is understood that we are talking about sinusoidal waveforms only:

$$V_0 = I_0 X_C, \tag{2.9}$$

---

[1] Of course, this is only an idealized approximation, since in any real circuit there is at least the resistance of the wires, and in practice any signal source has some internal resistance.

where $I_0$ is the amplitude of the current sine wave and $X_C \equiv 1/\omega C$ is the *capacitive reactance* of the capacitor. The reactance is thus the effective resistance of the capacitor. Note that it is frequency-dependent, in keeping with our intuition that for DC a capacitor should look like an open circuit (infinite resistance), while at high frequency it should approach a short circuit (zero resistance).

---

For completeness, we mention here the inductive reactance $X_L \equiv \omega L$, where $L$ is the *inductance* of an inductor. Inductors are coils of wire and satisfy the equation

$$V = L\frac{\mathrm{d}I}{\mathrm{d}t}. \tag{2.10}$$

Just as capacitors often employ a dielectric, inductors are often wound on a ferrite core to increase their inductance. Note that the inductor equation relates the voltage across an inductor to the *derivative* of the current through it, while the capacitor equation (Eq. 2.1) relates the voltage across a capacitor to the *integral* of the current. Thus, where the current through a capacitor *leads* the voltage across it, the current through an inductor *lags* the voltage across it by 90°. With respect to its function in a circuit, an inductor can thus be thought of as the opposite of a capacitor. Whereas capacitors are relatively small, light, cheap, and have negligible resistance, inductors tend to be large, heavy, expensive, and have appreciable resistance. Nevertheless, they find important use in filtering applications, e.g. bandpass filters, crossover circuits for hi-fi speakers, radio-frequency circuits, and so forth. In the interests of time we omit inductor exercises from our course, but if you understand capacitors you will have very little difficulty in applying inductors.

---

### 2.1.2 Types and values of capacitors

For some reason the various manufacturers' conventions for marking capacitors are particularly confusing – probably it has to do with the fact that many small-value capacitors are physically too small to permit much printing on them. Some common sense is required. Keep in mind that 1 farad is a huge unit! Most capacitors are in the picofarad and microfarad ranges, and these are the two commonly used units. A physically large capacitor that says '10M' on it is usually 10 microfarads, *not* 10 millifarads (for some reason, most manufacturers don't want to print Greek letters, so they use 'M' instead of 'μ'). A 10 millifarad cap would be labeled '10000M'. A small capacitor that says just '10' on it is 10 picofarads. The other important number is the maximum operating voltage, which is usually printed on the capacitor if there is room.

Some small capacitors are labeled like resistors, either with a color code or with numbers that mean the same thing. The first digit of this capacitance code is the tens, the second is the ones, the third is the power of 10, and the units are picofarads. This is sometimes ambiguous – for example, a capacitor that says '470' could be 470 pF or $47 \times 10^0 = 47$ pF! Usually the clue is the presence of a letter, following the capacitance code, that indicates the tolerance – J for $\pm 5\%$, K for $\pm 10\%$, M for $\pm 20\%$, etc. – so that '470 K' means 47 pF $\pm$ 10%, whereas just 470 means 470 pF! Note that there is no ambiguity if it says 471 – since normal capacitors are not manufactured with enough precision to warrant a third significant digit, the '1' must be the power of ten. When in doubt, you can always check it out by putting it in an $RC$ circuit with a known $R$ value and measuring the time constant (see below), or by plugging it into a capacitance meter, if you have one.

## 2.2  Review of current, voltage, and power

Before we get started on $RC$ circuits, let us briefly review power dissipation and component ratings – you need to understand these to avoid damaging components.

Voltage is related to potential-energy difference. The voltage drop across any circuit element is directly proportional to the change in energy of a charge as it traverses the circuit element. Specifically,

1 volt = 1 joule/coulomb.

The potential energy (with respect to some reference point) is equal to the voltage multiplied by the charge.

Current refers to the motion of charges. The current through a given surface (e.g. the cross-section of a wire) is defined as the net charge passing through that surface per unit time. The unit for current is the ampere:

1 ampere = 1 coulomb/second.

The product of voltage and current has units of joules/second, otherwise known as watts.

If the voltage drop across a circuit element equals the change in potential energy per unit charge, and the current equals the amount of charge moving through the element per unit time, then their product equals the power released within the device! The power dissipated within any device is given by

$$P = IV. \tag{2.11}$$

For resistive elements (or when an effective resistance can be defined), Eq. 2.11 can be combined with Ohm's law to give:

$$P = IV = I^2R = V^2/R. \tag{2.12}$$

Resistors, diodes, transistors, relays, integrated-circuit chips, etc., are rated (in part) by their maximum allowed power. Exceeding these ratings can have disastrous effects on your circuit, and may even cause a fire! To illustrate this point, our first exercise will deliberately lead to the destruction of a carbon-film resistor.

### 2.2.1 Destructive demonstration of resistor power rating

**Caution: In the following exercise, care must be taken to prevent burns.**

The resistor in the following exercise will become **very hot** and may even catch fire (briefly). Keep the body of the resistor well above the breadboard. Do **not** touch the resistor with your fingers. Remove the destroyed resistor using pliers or a similar tool.

Be sure that the power is turned off, and construct the circuit shown in Fig. 2.2 using a $\frac{1}{4}$ watt carbon-film resistor.

▷ Turn on the power and observe the effect on the resistor. Be sure to turn off the power as soon as the resistor begins to smoke. Record your observations and comments.

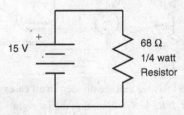

**Fig. 2.2.** This circuit can be used to demonstrate destructive power loading. Note that the resistor will heat up rapidly.

▷ Calculate the power that was dissipated by the resistor before it burned out. What is the minimum resistor value that can be safely used in this circuit? (Assume that only $\frac{1}{4}$ watt resistors are available.)

▷ Calculate the current that flowed through the resistor (before it burned out).

Note that even though the voltage was low and the current was well under 1 A, damage was nevertheless done! Because your body's resistance is large, low voltages can't give you a shock, but in the wrong circumstances they can still cause trouble. The key to safe work in electronics is always to estimate power dissipations in components *before* turning on the power, and to make sure you are not exceeding the ratings.

## 2.3 Potentiometer as voltage divider

The voltage-divider idea is very useful in analyzing almost all circuits, so you will need to become thoroughly familiar with it. A resistive voltage divider is simply two resistors in series (see Fig. 2.3). A voltage difference, $V_{in}$, is applied across the two, and a smaller voltage, $V_{out}$, results at the junction between them. A potentiometer can be used as a variable voltage divider, and you will now try this out using the breadboard's 10 k pot.

*Warning:* You can easily burn out the pot in this exercise if you are not careful!

1. Turn off the power before hooking up the circuit!
2. If you accidentally connect the pot's slider to ground while one end is connected to the supply voltage (or vice versa) you can easily burn out

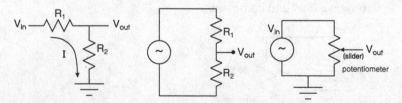

**Fig. 2.3.** Three schematics representing a resistive voltage divider. In all cases you can show using Ohm's law that $V_{out} = V_{in} R_2/(R_1 + R_2)$. Note that the far right representation is implemented using a potentiometer. In this case, the output voltage is variable and ranges between ground and $V_{in}$ (depending on the position of the slider).

the pot – briefly explain why this is true. (Hint: how much power can be dissipated in the pot in such a situation?)

3. If you connect the multimeter on a current or resistance setting between the slider and some other point in the circuit while the circuit is powered, you can easily burn out the pot, since on these settings a meter can act as a low impedance (short circuit).

### 2.3.1 DC voltage divider

First, use the ohmmeter setting of your multimeter to measure the resistance between the slider and each end of the 10 k pot.[2] Turn the knob to set the slider exactly half-way between the ends, using the meter to tell you when you get there.

Next, use the voltmeter setting to adjust the variable positive power supply to +10 V. Then, turn off the power, connect one end of the pot to ground, and connect the other end to +10 V. Connect the meter (on the voltage setting) between the slider and ground. Double-check your connections.

▷ Without moving the slider or changing the supply voltage, turn on the power, and measure the voltage between the slider and ground.

▷ What are the values of $R_1$ and $R_2$? Using the voltage-divider equation

$$V_{\text{out}} = V_{\text{in}} \left( \frac{R_2}{R_1 + R_2} \right), \tag{2.13}$$

explain why the predicted output voltage is +5 V. How close to this prediction is your measured voltage? What is the percentage error?

### 2.3.2 AC voltage divider

Now verify that a resistive voltage divider works the same way for AC as for DC. Apply a sinusoidal signal from the function generator to the pot in place of the +10 V DC.

▷ Look at the function generator's output signal with the scope and measure its peak-to-peak voltage, amplitude, and r.m.s. voltage. (The scope's MEASURE menu is useful here.) Look at the signal at the pot's center tap – what are the peak-to-peak, amplitude, and r.m.s. values there? How does the pot's voltage-division ratio, $R_2/(R_1 + R_2)$, compare for DC and AC?

---

[2] The 10 k pot is located near the center of the bottom edge of the PB-503 breadboard and is adjusted by means of a large black knob. If you don't have a PB-503, find a 10 k pot on your breadboard if it has one; otherwise you will have to use a separate 10 k pot.

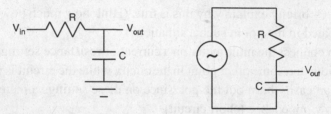

**Fig. 2.4.** The voltage-divider concept works perfectly well for *RC* circuits. This circuit is also known as a low-pass filter, or as a voltage integrator.

## 2.4 *RC* circuit

Now hook up a 10 k resistor and a 0.01 μF ceramic capacitor in series. Ground one end of the capacitor, connect the other end to the resistor, and connect the other end of the resistor to a 500 Hz square wave from the function generator (see Fig. 2.4). Display the input signal (output of the function generator) on channel 1 of the scope, and the output signal (at the junction of the resistor and capacitor) on channel 2.

▷ What are the amplitudes of the input and output waveforms?

▷ Sketch the output waveform. You may recall from your general physics that it obeys the equation

$$V(t) = V_0 e^{-t/RC}, \tag{2.14}$$

where $t = 0$ corresponds to a rising or falling edge of the square wave.

▷ Show that the time to fall to 37% of the peak value (i.e. $V(t) = 0.37\,V_0$) is the 'time constant', $RC$, and determine $RC$ using the scope – you should find your oscilloscope's 'cursor' feature useful here. (Be sure to set the time and voltage scales sufficiently sensitive to yield an accurate measurement – what settings should you use, and why?)

▷ Based on the nominal component values, what do you predict for $RC$? Is your measurement consistent with this prediction? What are the tolerances of the components you are using – does this explain any discrepancy?

## 2.5 *RC* circuit as integrator

Now switch the function-generator frequency to 50 kHz. Observe what happens to the output waveform's shape and amplitude. This can be explained quantitatively from Eq. 2.1

$$V_{out}(t) = \frac{Q(t)}{C} \tag{2.15}$$

$$= \frac{1}{C} \int_0^t I(t)dt \tag{2.16}$$

$$= \frac{1}{C} \int_0^t \frac{V_{in} - V_{out}}{R} dt \tag{2.17}$$

$$\approx \frac{1}{RC} \int_0^t V_{in}dt, \tag{2.18}$$

where $V_{out}$ is the voltage across the capacitor, and the approximation is valid as long as $V_{out} \ll V_{in}$.

▷ What does Eq. 2.18 predict when $V_{in}$ is a constant, as it is during half of each period of the square wave? Carefully measure and sketch the output waveform. Compare your observations with your expectations based on Eq. 2.18 and explain your results.

▷ What output amplitude would you expect at 25 kHz? Change the input frequency and see if your prediction is correct.

▷ At approximately what frequency will Eq. 2.18 cease to predict the output waveform accurately? Change the input frequency and test your prediction.

## 2.6 Low-pass filter

Now switch from a 50 kHz square wave to a 50 kHz sine wave. Since Eq. 2.18 should still apply, the output waveform should be the integral of a sine wave, i.e. a cosine wave.

▷ What does this imply about the *phase shift* between input and output? Measure the phase shift: 360° multiplied by the time $\Delta t$ between the zero crossing of the input signal and the zero crossing of the output signal, divided by the period, or

$$\phi = 360° \frac{\Delta t}{T}. \tag{2.19}$$

(The cursors are useful here.)

▷ Is the measured phase shift consistent with your prediction? Does the voltage across the capacitor lag or lead the current through it? Explain.

The other way to analyze this circuit is as an AC voltage divider, using Eq. 2.13 with $R_2$ replaced by the reactance of the capacitor, $X_C$, and $R_1 + R_2$ replaced by the total impedance, $Z$, of the resistor and capacitor in series. Since the voltage across the capacitor is 90° out of phase with the current through the resistor, these add in Pythagorean fashion:

$$V_{out} = V_{in}\frac{X_C}{Z} \tag{2.20}$$

$$= V_{in}\frac{X_C}{\sqrt{R^2 + X_C^2}} \tag{2.21}$$

$$= V_{in}\frac{1}{\sqrt{1 + (\omega RC)^2}}. \tag{2.22}$$

We see that the attenuation ($V_{out}/V_{in}$) depends on the frequency.

▷ What attenuation do you observe at 50 kHz? Calculate what you expect, and compare.

At the *breakpoint* or *half-power* frequency $f_0$, $\omega RC = 1$, and thus the attenuation is $1/\sqrt{2} = 0.707$. This is also referred to as the $-3$ dB point, since it is the frequency at which the output voltage is attenuated by 3 dB. The breakpoint is a convenient way to parametrize simple filters. (The decibel is a logarithmic measure of the ratio of two signals:

$$\text{number of dB} = 20\log\frac{A_2}{A_1},$$

where $A_1$ is the amplitude of the first signal and $A_2$ is the amplitude of the second; in this case $A_1 = V_{in}$ is the amplitude of the output signal and $A_2 = V_{out}$ is the amplitude of the input signal.)

▷ By varying $f$ until the output amplitude is 70.7% of the input amplitude, measure $f_0$. Calculate what you expect, and compare with your measurement.

▷ What are the attenuation and phase shift at low frequency, say 50 Hz? Compare with the predictions of Eq. 2.22. Compare the phase shifts with

$$\phi = \arctan\frac{R}{X_C}. \tag{2.23}$$

The phase shift at low frequency is easy to understand: in the limit of DC the capacitor must act like an open circuit, i.e. infinite impedance, and thus does not affect the output signal. Conversely, in the high-frequency limit the capacitor must look like a short circuit to ground, so the output signal goes to zero and the phase shift becomes dominated by the capacitor.

## 2.7 *RC* circuit as differentiator

Now interchange the capacitor and resistor so that the input signal is applied at the capacitor (see Fig. 2.5). Drive the circuit with a 50 Hz square wave.

▷ What waveform do you see at the output? What are the input and output amplitudes?

You can think of the shape of the output in terms of the exponential $RC$ charging/discharging curve, with $f \ll 1/RC$, or you can think of it as an approximation to the derivative of the input signal. Mathematically, the derivative of an ideal square wave would be infinite at the voltage steps and zero in between, but of course an electrical signal can never be infinite! In this circuit the voltage spikes are limited in size to twice the input amplitude. Using Eq. 2.16,

$$V_{\text{out}} = IR \tag{2.24}$$

$$= R\frac{dQ}{dt} \tag{2.25}$$

$$= RC\frac{d(V_{\text{in}} - V_{\text{out}})}{dt} \tag{2.26}$$

$$\approx RC\frac{dV_{\text{in}}}{dt}, \tag{2.27}$$

where the approximation is again valid when $V_{\text{out}} \ll V_{\text{in}}$. So, indeed, the circuit puts out an approximation to the time derivative of the input signal. You can see why the approximation of Eq. 2.27 breaks down in the case of a square wave, since at the rising and falling edges of the square wave $V_{\text{out}} > V_{\text{in}}$.

▷ What does Eq. 2.27 imply if the input is a triangle wave? Try it out and compare quantitatively with what you expect.

▷ What does Eq. 2.27 imply if the input is a sine wave? Try it out and compare quantitatively with what you expect. Sketch the output waveform.

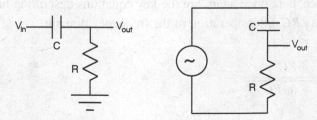

**Fig. 2.5.** High-pass filter or voltage differentiator.

If you are surprised at all the wiggles on the output signal, you can verify that they are real (as opposed to noise) using the signal-averaging feature of the scope's ACQUIRE menu. You've discovered a poorly kept secret of function-generator design! The sine waveform is rather difficult to generate, and most function generators actually use an approximation to it that is piecewise-linear around the peaks and valleys. The derivative of a piecewise-linear function is a series of steps and plateaus.

## 2.8 High-pass filter

▷ What attenuation and phase shift do you observe with a 50 Hz sine wave as input?
▷ What about with a 50 kHz sine wave?
▷ Why do these phase shifts make sense?
▷ Should the breakpoint frequency be any different in this configuration than in the low-pass filter? Check it and make sure. Compare your measurements with

$$V_{\text{out}} = V_{\text{in}} \frac{R}{Z} \tag{2.28}$$

$$= V_{\text{in}} \frac{R}{\sqrt{R^2 + X_C^2}} \tag{2.29}$$

$$= V_{\text{in}} \frac{\omega RC}{\sqrt{1 + (\omega RC)^2}}. \tag{2.30}$$

▷ Show that well below the breakpoint frequency, Eq. 2.30 predicts that the output amplitude should increase linearly with frequency. Take a few measurements to demonstrate that this prediction is correct.

## 2.9 Summary of high- and low-pass filters

For reference, here once again are the key equations describing high-pass and low-pass $RC$-filter operation in the frequency domain.
High-pass:

$$\frac{V_{\text{out}}}{V_{\text{in}}} = \frac{\omega RC}{\sqrt{1 + (\omega RC)^2}}, \tag{2.31}$$

$$\phi = \arctan \frac{1}{\omega RC}. \tag{2.32}$$

Low-pass:

$$\frac{V_{\text{out}}}{V_{\text{in}}} = \frac{1}{\sqrt{1 + (\omega RC)^2}},\qquad\qquad (2.33)$$

$$\phi = \arctan \omega RC. \qquad\qquad (2.34)$$

Note, as expected, that at high frequency, the voltage-division ratio goes to unity in the first case and to zero in the second case, while the phase shift goes to zero in the first case and to 90° in the second. Also, in both cases, the breakpoint frequency is the point at which the two terms in the

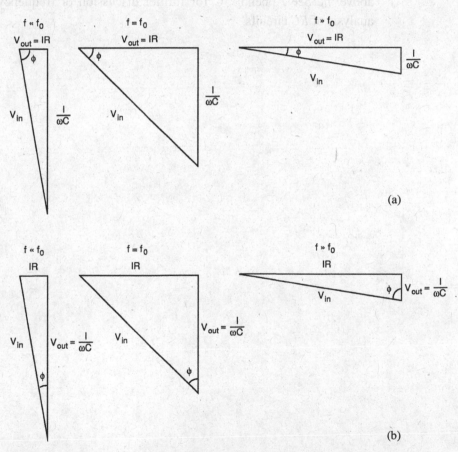

(a)

(b)

**Fig. 2.6.** Right triangles depicting the relationships among input voltages (always represented by the hypoteneuse of the triangle) and capacitor and resistor voltages for (a) high-pass and (b) low-pass *RC* filters. In each case, the center diagram shows the isosceles triangle representing the case $f = f_0$; the triangles on the left are for a frequency well below $f_0$; and those on the right for a frequency well above $f_0$.

square-root are equal:

$$\omega_0 RC = 1 \qquad\qquad (2.35)$$

$$f_0 = \frac{\omega_0}{2\pi} \qquad\qquad (2.36)$$

$$= \frac{1}{2\pi RC}. \qquad\qquad (2.37)$$

These relationships are all illustrated in Fig. 2.6, which shows, for low- and high-pass filters, how the phase and magnitudes of the voltages across the resistor and capacitor are related, for a fixed input voltage, in three frequency regimes: $f = f_0$ and frequencies that are well below and well above $f_0$. See Appendix C for further discussion of frequency-domain analysis of $RC$ circuits.

# 3 Diodes

In this chapter we will explore semiconductor diodes and some circuits using them. We've seen that resistors have a simple linear relationship between the voltage across them and the current through them (Ohm's law). On the other hand,

- diodes have an exponential relationship between current and voltage.

Mathematically this may seem much more complicated than Ohm's law, but we think you'll agree that the idea as just stated is simple enough – it just takes some getting used to! As we'll see, an important consequence of the exponential characteristic is that diodes conduct much more readily in one direction than in the other. This makes them ideally suited for *rectification*: the conversion of AC into DC.

## Apparatus required

Breadboard, oscilloscope, one or two multimeters, one 1N914 (or similar) silicon signal diode, one 1N4001 (or similar) 1 A silicon rectifier diode, one 100 $\Omega$ and one 10 k $\frac{1}{4}$ W resistor, one 1 k 2 W resistor, power transformer with 12.6 V r.m.s. output on each side of the center tap, one diode bridge element, one 100 $\mu$F electrolytic capacitor, and one 1000 $\mu$F electrolytic capacitor.

## 3.1 Semiconductor basics

Current will flow through a material provided that there are charge carriers free to move and an electric field to move them. Conductors (such as copper) have lots of charge carriers (electrons) ready to move in response to the slightest electric field. Insulators (such as diamond) possess very few free charge carriers – all the electrons are tightly bound to the crystal lattice, so that, even in the presence of a strong electric field, no current flows. Semiconductors (such as silicon and germanium) are somewhere in between. The conductivity of a semiconductor can be enhanced through doping, the deliberate inclusion of impurities within the semiconductor lattice.

Silicon, for example, has four valence electrons, which are used to make covalent bonds with neighboring silicon atoms. Phosphorus has five valence electrons, and boron has three. In a silicon crystal, if a silicon atom is replaced with a 'donor' material, such as phosphorus, an extra valence electron becomes available that is loosely bound to the lattice. If an 'acceptor' material such as boron is substituted for silicon, a 'hole' appears in the electron structure of the lattice. Doping silicon with a donor material creates an 'N-type' semiconductor, whereas doping with an acceptor creates a 'P-type' semiconductor. (Note that, despite their names, these doped semiconductors are electrically neutral: the 'extra' electrons in N-type material are compensated for by the additional protons in the atomic nuclei of the donors, while the 'missing' electrons in P-type are compensated for by the missing protons in the acceptor nuclei.)

The 'extra' electrons within N-type material can move under the influence of an electric field. Thus, the dominant charge carriers are electrons; i.e., N-type material has negative charge carriers.

For P-type material, electrons from neighboring atoms can jump into the holes, moving the holes from one place to another. The holes can migrate in the direction of an electric field. The charge motion is thus due to the motion of the holes, i.e., P-type material has positive charge carriers.

If a junction between P-type and N-type semiconductor material is created within a single crystal, in such a way that the crystalline structure is preserved across the junction, the result is a *junction diode*. Electrons from the N-region migrate across the junction into the P-region, filling holes as they go. This creates a net charge build-up around the junction (see Fig. 3.1) – positive in the N-region and negative in the P-region – leading to an internal electric field as shown. Once the holes are filled, the junction region becomes devoid of charge carriers and thus acts as an insulator, preventing further current flow.

If an external field is applied in the *same* direction as the internal field, the 'depletion region' (region around the junction devoid of charge carriers) increases in size, so current does not flow. On the other hand, if an external field is applied *opposite* to the internal field, free charge carriers flow toward the junction. Electrons flow into the N-type material from the metal contact

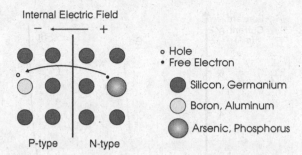

**Fig. 3.1.** Representation of a junction between P-type and N-type semiconductor material. Free electrons from the N-region will migrate into the P-region, combining with holes.

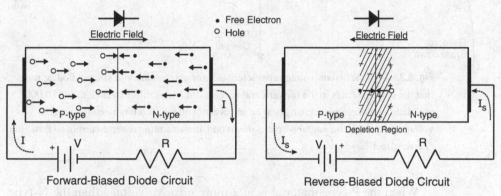

**Fig. 3.2.** Diode circuit symbol and biasing.

(see Fig. 3.2). New holes are created within the P-material as electrons jump from the semiconductor to the metal contacts. At the junction, the holes from the P-type material meet electrons from the N-type material and combine. A PN junction thus allows current to flow easily in one direction but blocks current flow in the reverse direction.

For such a diode the current $I$ flowing through the device is given approximately by

$$I = I_s(e^{eV/nkT} - 1) \tag{3.1}$$

where $I_s$ (sometimes called $I_R$ or $I_0$) is the 'reverse saturation current', $e$ is the electron charge, $V$ is the voltage across the junction, $n$ is an empirical constant between 1 and 2, $k$ is Boltzmann's constant, and $T$ is the junction temperature in kelvin. For simplicity, we'll assume for now that $n = 1$. This dependence of current on voltage is illustrated in Fig. 3.3.

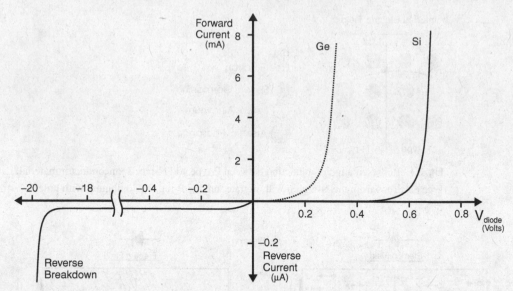

**Fig. 3.3.** Typical current–voltage characteristics for germanium and silicon diodes; note that the current scales in the forward and reverse directions differ by a factor of 10 000, and that the voltage scale changes at large reverse voltage. If a large enough reverse voltage is applied, the junction breaks down and allows a large reverse current to flow (the 'Zener effect').

When the P-type material is at a more *positive* voltage than the N-type material, the diode is said to be 'forward-biased'; this corresponds to $V > 0$ in Fig. 3.3. When the P-type material is more *negative* than the N-type material, the diode is said to be 'reverse-biased'; this corresponds to $V < 0$ in Fig. 3.3.

## Some useful approximations

In the forward-biased case, when $V$ is greater than $\approx 100$ mV or so, the '1' in Eq. 3.1 becomes negligible compared with the exponential term, and

$$I \approx I_s e^{eV/kT}. \tag{3.2}$$

When the diode is reverse-biased and $|V|$ is greater than $\approx 100$ mV or so, the exponential term is negligible, and the reverse current is almost constant, with

$$I \approx -I_s. \tag{3.3}$$

## 3.2 Types of diodes

In addition to standard junction diodes, light-emitting diodes (LEDs), Schottky diodes, and Zener diodes are also common. LEDs are junction diodes typically made from gallium arsenide phosphide (GaAsP). They act very much like silicon junction diodes except that they emit light when conducting forward current and have forward voltage drops about twice as large as silicon diodes. Infrared, red, orange, yellow, green, and blue LEDs are commercially available.

Zener diodes are manufactured with controlled reverse-breakdown properties. Their forward characteristics are similar to those of junction diodes; however, Zener diodes are used in reverse-biased mode. While reverse breakdown typically destroys a standard junction diode, Zener diodes are designed to operate at and around their reverse-breakdown ('Zener') voltage. The Zener voltage is determined during the manufacturing process by adjusting the semiconductor doping. Typical Zener voltages range from 3.3 to 75 volts.

Schottky diodes are manufactured by bonding a metal conductor to an N-type semiconductor. Electrons from the N-type material migrate into the metal. This migration creates a potential barrier across the boundary, which then behaves in a similar fashion to a PN junction. In general, Schottky diodes are used in applications requiring high speed and low capacitance.

Physically, most diodes look like a little cylinder with wires sticking out the two ends (Fig. 3.4). To distinguish the ends from each other, the manufacturer often prints the diode circuit symbol on the diode body. Alternatively, sometimes a ring is marked around the body close to the 'cathode' (the N-type end), to distinguish it from the 'anode' (the P-type end). Diodes are manufactured with specified values for maximum current, forward voltage drop, leakage current (reverse saturation current), reverse-breakdown voltage, and switching speed (time required for the diode to

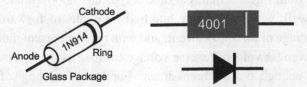

**Fig. 3.4.** Representation of physical diodes along with the symbols used in circuit diagrams.

**Table 3.1.** A diverse selection of diodes is commercially available, of which a tiny sampling is given here.

Diodes are commonly rated by their switching speed, maximum power dissipation, maximum forward current, maximum forward voltage at a specified forward current, and reverse-breakdown voltage. The junction capacitance is sometimes listed as well.

| Diode | Type | $P_{max}$ (W) | $I_{F_{max}}$ (A) | $V_{F_{max}}$ @ $I_F$ (V @ A) | $V_{BR(R)}$ (V) | $C$ (pF) |
|-------|------|------|------|------|------|------|
| 1N914 | small signal | 0.5 | 0.3 | 1.0 @ 0.01 | 75 | 4.0 |
| 1N4001 | rectifier | | 1.0 | 1.1 @ 1.0 | 50 | 8.0 |
| 1N4004 | rectifier | | 1.0 | 1.1 @ 1.0 | 400 | 8.0 |
| 1N5402 | rectifier | | 3.0 | 1.1 @ 3.0 | 200 | 40 |
| FR601 | fast rectifier | | 6.0 | 1.3 @ 6.0 | 50 | 200 |
| MBD301 | Schottky | 0.28 | 0.1 | 0.6 @ 0.01 | 30 | 1.0 |
| 1N4733A | Zener | 1.0 | | 1.2 @ 0.2 | 5.1 | |

switch from forward to reverse bias or vice versa). A few examples are given in Table 3.1.

## 3.3 Rectification

A *rectifier* is a device that converts AC to DC by blocking the flow of current in one direction. Rectification used to be almost the exclusive province of vacuum tubes, with the exception of the 'detector crystals' (naturally occurring semiconductor diodes) used in crystal sets in the early days of radio. Nowadays, semiconductor diodes are universally used for the purpose.

An ideal rectifier would offer *zero* resistance when forward-biased, i.e., the voltage across it would be zero, independent of the amount of forward current flow. It would offer *infinite* resistance when reverse-biased, i.e., no current would flow, regardless of the size of the applied reverse voltage. You can see from Fig. 3.3 that real diodes typically approximate the ideal reasonably well, with only a few hundred millivolts of forward voltage over a wide range of forward current, and with reverse current measured in nanoamps, even for volts of reverse voltage (as long as the reverse voltage is kept small enough to avoid breakdown). For the typical range of currents encountered in most electronic circuits (a few to hundreds of milliamps), a handy approximation is that a forward-biased germanium diode has about a 300 mV voltage drop across it, while a forward-biased silicon diode has about a 600 mV drop.

## 3.4 Diode action – a more sophisticated view

If we want, we can think of a diode as a resistor whose resistance depends on the current flowing through it, i.e., the resistance is *dynamic* rather than having a constant (or *static*) value. While the static resistance of a device is the voltage across the device divided by the current through it ($R = V/I$, Ohm's law), the dynamic resistance is the slope $dV/dI$ of the $V-I$ curve at any point. You can see that, for a resistor, the static and dynamic resistances are the same, but for a device whose $V-I$ curve is nonlinear, they become different. Note in particular that a nonlinear device does not have a static resistance, since $V/I$ is not constant.

We can find the dynamic resistance of a diode by differentiating Eq. 3.1:

$$\frac{dI}{dV} = \frac{e}{kT}I_s e^{eV/kT};\tag{3.4}$$

thus, the dynamic resistance is

$$\frac{dV}{dI} = \frac{kT/e}{I_s e^{eV/kT}}\tag{3.5}$$

$$\approx \frac{kT/e}{I},\tag{3.6}$$

where the approximation is valid for forward-biasing such that the '1' in Eq. 3.1 is negligible. Note that at $T = 300$ K (room temperature),

$$\frac{e}{kT} = \frac{1.6 \times 10^{-19}\,\text{C}}{(1.38 \times 10^{-23}\,\text{J/K}) \times 300\,\text{K}} = 39\,\text{V}^{-1} \approx \frac{1}{25\,\text{mV}},\tag{3.7}$$

so the dynamic resistance of a forward-biased diode can be simply approximated by

$$\frac{dV}{dI} \approx \frac{25\,\text{mV}}{I}.\tag{3.8}$$

These results – the exponential dependence of current on voltage and the consequent dynamic-resistance formula – are important to remember, as they also characterize the behavior of transistors. (This model of diode behavior neglects some features that matter in practice. For example, the semiconductor has some 'ohmic', or static, resistance that is in series with the junction resistance just described, so in practice the dynamic resistance is somewhat larger than that given by Eq. 3.8.)

## 3.5 Measuring the diode characteristic

As before, be careful not to burn out your breadboard's 1 k pot: hook up the following circuits with the power off, and double-check each circuit before powering it up.

▷ For the circuit shown in Fig. 3.5, estimate the maximum current in your circuit and the maximum power dissipation in the 100 $\Omega$ resistor – is the resistor's $\frac{1}{4}$ W rating safe for its worst-case power dissipation? **Caution:** These small diodes are easily damaged by overcurrent. To be on the safe side, do not let the forward current exceed 50 mA.

▷ With a 1N914 (or similar) silicon diode, forward-biased as shown in Fig. 3.5, increase the voltage across the diode starting from 0 V in steps of 100 mV and record the diode current in each case.

For the portion of the characteristic curve close to the origin, the microamp range of the multimeter will be required to measure the forward current. If you don't have a second meter available, you can use your oscilloscope to measure the diode voltage.

▷ Plot your results for $I$ vs. $V$ on a linear scale – do they seem qualitatively consistent with the functional shape of Eq. 3.1?

When unity is negligible in comparison with the exponential term of Eq. 3.1, Eq. 3.2 may be re-expressed as

$$\ln I \approx \ln I_s + \frac{e}{kT}V. \tag{3.9}$$

▷ Plot $\ln I$ vs. $V$. Is $\ln I$ approximately linear in $V$? How does its slope compare with $e/kT$? What value for $n$ (from Eq. 3.1) is implied by your measured slope?

▷ Now reverse the diode – for a reverse voltage of 5 V what reverse current do you observe? Is it consistent with the range of $I_s$ expected for a silicon diode?

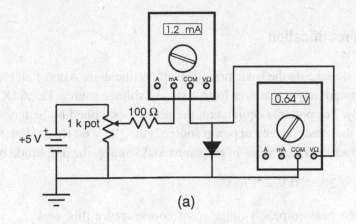

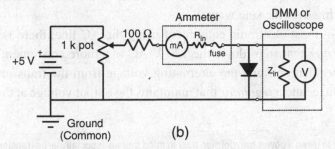

**Fig. 3.5.** (a) Measuring the forward characteristic of a diode. (b) When used to measure current, the DMM is equivalent to an ideal ammeter in series with a small input impedance. Most ammeters will also have a series fuse to protect the meter. When measuring voltage, the DMM or oscilloscope looks like an ideal voltmeter in parallel with a large input impedance.

If the reverse current seems to be much bigger than you expect, consider that you have a voltage-measuring device (a scope or voltmeter) in parallel with the diode (Fig. 3.5(b)).

▷ Disconnect the scope or meter – now how much current do you observe? Reconnect it and disconnect the diode – how much current flows with the scope or meter alone? What do you infer to be the input resistance of the scope or meter? Explain by applying Ohm's law to relate the voltage being measured to the current you observe.

Keep this experience in mind – it is often necessary to consider the effect of your measuring device on the circuit being studied. Can you see why an ideal voltmeter would have infinite resistance, while an ideal ammeter would have zero resistance?

## 3.6 Exploring rectification

Next we take up the basic principles of rectification. Almost all electronic equipment requires power from a steady voltage source, i.e., a DC power supply. For portable equipment, the power is supplied by batteries. However, the most convenient power source is the 120 V 60 Hz AC line.[1] (120 V is in fact the r.m.s. value of the sinusoidal voltage, the amplitude being

$$V_0 = \sqrt{2} \times 120 \text{ V} = 170 \text{ V}, \tag{3.10}$$

and the peak-to-peak voltage is, of course, twice this, or $V_{p-p} = 340$ V, as you can easily verify from the definition of the root-mean-square by integrating over the sine wave.)

Within most electronic equipment using the AC line, there is a *power transformer* that steps down the 120 V AC to a more convenient voltage, a *rectifier* that converts the alternating voltage from the transformer to a DC voltage, and a *regulator* that maintains the output voltage at the desired level.

> **Caution:** In using a power transformer, bear in mind that an especially large transient current sometimes flows when the line cord is first plugged in.

You will probably blow fewer fuses if you leave the power transformer plugged in at all times. Attach banana-plug leads to the transformer's secondary only after you are sure your circuit will not damage any of the equipment. Do not permit powered lines to dangle loosely; when reconfiguring your circuit, it is safest to disconnect the leads at the transformer, not at the breadboard.

▷ Set up the circuit shown in Fig. 3.6(a) using a 10 k resistor as the load, $R_L$. Observe the sinusoidal voltage waveform across $R_L$. Measure the amplitude $V_0$ and the r.m.s. voltage. Check the relation

$$\sqrt{2}V_{rms} = V_0. \tag{3.11}$$

You will probably find $V_{rms} > 25$ V. Since the windings of the transformer have some ohmic resistance, the transformer's output voltage depends on

---

[1] In North America, the supply voltage from a standard wall socket is 120 V, and the supply frequency is 60 Hz; the discussion is equally valid for other values, which may be substituted according to your local supply voltage and frequency.

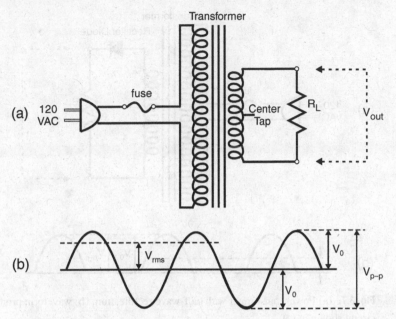

**Fig. 3.6.** (a) Power transformer supplies $V_{out} \approx 25$ V r.m.s.; (b) waveform produced by the circuit in (a).

the current drawn, and its 25.2 V r.m.s. nominal output voltage is for substantially higher current than is drawn by the 10 k load.

▷ Add a 1N4001 diode to give the half-wave rectifier of Fig. 3.7(a) with $R_L = 10$ k.

▷ Observe and record the voltage waveform. Measure the amplitude $V_0$ using the oscilloscope (due to the rectification it is now equal to the peak-to-peak voltage).

▷ Compare the amplitude of the half-rectified waveform with the amplitude of the unrectified waveform measured above. By how much has the amplitude decreased? Is this the amount you expect? Explain.

▷ Measure the average voltage $V_{av}$ across the load with a DC voltmeter. Check that for a half-wave rectifier

$$V_{av} = \frac{V_0}{\pi}. \tag{3.12}$$

▷ Add a filter capacitor in parallel with the load as shown in Fig. 3.8(a).

---

**Caution:** The 100 μF electrolytic capacitor is *polarized* – be careful not to hook it up backwards! The negative terminal should be labeled with a '–' sign.

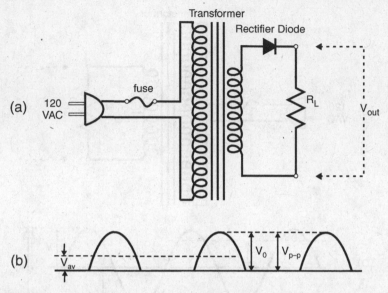

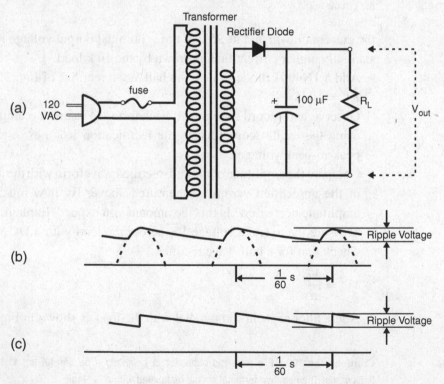

**Fig. 3.7.** (a) Power transformer with half-wave rectification; (b) waveform produced by circuit shown in (a).

**Fig. 3.8.** (a) Half-wave rectifier with filter capacitor; (b) waveform produced by circuit shown in (a); (c) simple approximation to waveform produced by circuit shown in (a).

▷ What is the voltage rating of your capacitor? Make sure it is sufficient for the voltage that will be applied!

▷ Observe and record the output-voltage waveform across the load resistance $R_L$ and measure the peak-to-peak 'ripple voltage' (i.e., the amount by which the output voltage is varying; see Fig. 3.8(b)).

Here is that rare situation – measuring accurately a small AC signal on top of a large DC offset – in which you should use the AC-coupling feature of the scope's vertical menu. If you are troubled with noise, you may want to trigger the scope on 'line' and employ signal averaging.

A simplified analysis approach for predicting the expected output waveform, illustrated in Fig. 3.8(c), is to assume that the capacitor charges up to the peak voltage instantaneously and discharges at a uniform rate $dQ/dt$, equal to the average load current. The average current through the load can be determined using the known resistance $R_L$ and the average DC voltage across it as measured with a voltmeter. Using these assumptions, and the fundamental capacitor equation $Q = CV$,

▷ Calculate the output-voltage droop in each cycle, and compare with the peak-to-peak ripple voltage as measured. Is the observed percentage discrepancy within the tolerances of your components? Explain.

▷ Replace the 100 μF electrolytic capacitor with a 1000 μF capacitor. Do you expect the ripple voltage to increase or decrease? Explain.

▷ Measure the ripple voltage and compare with your expectations.

Full-wave rectification should decrease the ripple by about a factor of two. This can be accomplished using two diodes and the transformer's center tap or by using a diode bridge. Diode-bridge rectifiers are available as a single encapsulated unit, making the bridge-rectifier approach particularly convenient. These bridges have four diodes within. The terminals are labeled: '$\sim$' marks the two terminals that should be connected to the transformer secondary, and '$+$' and '$-$' denote the positive and negative outputs (see Fig. 3.9).

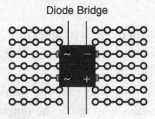

**Fig. 3.9.** An example of how to insert a diode bridge into a breadboard.

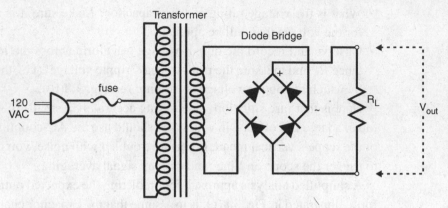

**Fig. 3.10.** Full-wave rectification using a diode bridge.

**Caution:** A defective bridge element can blow the power transformer fuse – check it before placing it in service. It should show essentially infinite resistance between the terminals marked '~'. When using an ohmmeter to check the resistance, remember to measure it for both orientations of the terminals – since you are dealing with diodes, the resistance could be different in each direction.

▷ Set up the bridge rectifier circuit of Fig. 3.10 with $R_L = 10$ k. (Insert the rectifier package straddling the central groove of a breadboard socket unit, with the long dimension of the package running along the groove, as shown in Fig. 3.9.) Four rectifier diodes can be used if a bridge rectifier is not available. As before, observe and record the voltage waveform across $R_L$.

▷ Add a filter capacitor to the full-wave rectifier as shown in Fig. 3.11(a) – again, be careful not to connect the capacitor backwards. This form of power supply is very common. Measure the average output voltage across the 10 k load. Record the peak-to-peak ripple voltage.

▷ Repeat these measurements for $R_L = 1$ k. **Caution:** What power rating must the 1 k resistor have, and why? Use socket adapters to handle the fat leads of the 2 W resistor.

▷ Repeat the ripple voltage calculations for these two values of $R_L$, keeping in mind that the filter-capacitor discharge time is now one-half of the AC cycle.

To make this circuit into a 'complete' power supply, one would want to *regulate* the output, that is, employ feedback to make the output voltage and ripple less dependent on the load resistance. This could be done using Zener diodes, but a more effective technique is a transistorized regulating

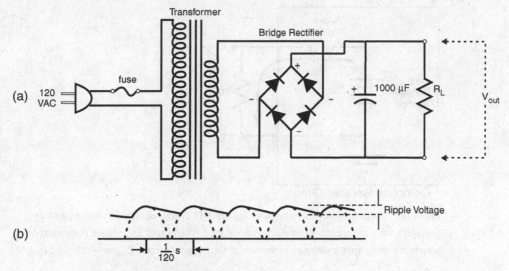

**Fig. 3.11.** (a) Full-wave rectification with filter capacitor; (b) waveform produced by circuit shown in (a).

circuit. Integrated three-terminal voltage regulators (such as the 7800 and 7900 series) have made this particularly simple.

## 3.7 Input and output impedance

Input and output impedance are key ideas that are used all the time in analyzing circuits. You've already encountered the input impedance of the scope or voltmeter in section 3.5.

A good way to think about the effect of an instrument on the circuit to which it is connected is via the instrument's Thévenin equivalent. The Thévenin equivalent of the multimeter (when set to measure voltage) is a large resistor in parallel with an ideal voltmeter (Fig. 3.5). In practice, an input also has some small capacitance and inductance and hence is more completely characterized by its impedance vs. frequency, which takes into account both the resistance and the capacitive and inductive reactances.

Outputs can also be characterized by their impedance (see Fig. 3.12). You've already taken data that determine the output impedance of your filtered full-wave rectifier circuit.

▷ From

$$Z_{\text{out}} = -\Delta V_{\text{out}}/\Delta I_{\text{out}},\qquad(3.13)$$

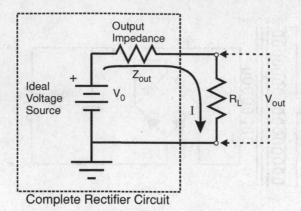

Complete Rectifier Circuit

**Fig. 3.12.** A rectifier circuit can be modeled (as a Thévenin equivalent) using an ideal voltage source in series with an output impedance. The output impedance is measured by observing the output voltage as a function of the output current: $Z_{out} = -\Delta V_{out}/\Delta I_{out}$.

and your data on $V_{out}$ vs. $R_L$, compute the circuit's output impedance, $Z_{out}$, in ohms.

▷ As another example, determine the output impedance of your bread-board's function generator, by measuring its sine-wave output amplitude, first with no load, and then with a load resistance of 1 k to ground.

If the output impedance has negligible frequency dependence, it can be approximated as a pure resistance, in which case the function generator's Thévenin-equivalent circuit consists of an ideal AC voltage source (one having zero internal resistance) in series with a single resistor.

▷ Check the function generator's $Z_{out}$ both at low and high frequencies (say 50 Hz and 50 kHz) – do you observe any appreciable frequency dependence?

▷ Sketch schematic diagrams (with component values labeled) of the Thévenin-equivalent circuits of your voltmeter, full-wave-rectified power supply, and function generator.

# 4 Bipolar transistors

Invented in 1947, transistors (and *integrated circuits* made from them) have been the basis for the explosive proliferation of electronic devices that revolutionized so much of life in the latter half of the twentieth century. Although discrete (i.e. individually packaged) transistors are now used mainly in special situations (e.g. where high power or speed is required), since transistors form the basis of a large class of integrated circuits, an understanding of how they work remains valuable. This will be the subject of the next few chapters. This chapter will introduce you to some basic bipolar-junction-transistor circuits.

### Apparatus required

Breadboard, oscilloscope, two multimeters, 2N3904 and 2N3906 transistors, red light-emitting diode (LED), 1N914 (or similar) silicon signal diode, two 330 $\Omega$, two 10 k, and one each of 100 $\Omega$, 1 k, 3.3 k, 22 k, and 100 k $\frac{1}{4}$ W resistors, 1 $\mu$F capacitor.

## 4.1 Bipolar-junction-transistor basics

Why and how transistors work is a bit subtle and can easily confuse the beginning student, but it is something you must master. Study the following description carefully, and compare it with the descriptions in other books. You may also want to re-read both our description and others after you've had some experience building and analyzing transistor circuits. (If you want more of the background detail on semiconductor physics, good places to look are Simpson's *Introductory Electronics for Scientists and Engineers*, or any textbook on modern physics.)

A bipolar junction transistor consists of two PN junctions sandwiched very close together within a single crystal of semiconductor (Fig. 4.1(a)).

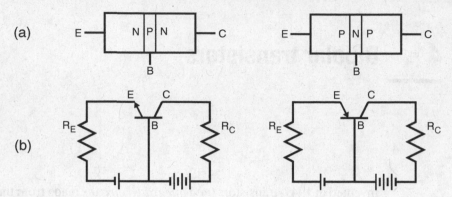

**Fig. 4.1.** (a) Construction and (b) circuit symbols and biasing examples for NPN and PNP junction transistors.

The region common to the two junctions, called the *base*, may be of either N-type or P-type material. This thin region is surrounded by material of the opposite type, in regions known as the *emitter* and *collector*. Wire leads are attached to the three regions.

The circuit symbols for NPN and PNP junction transistors are shown in Fig. 4.1(b). Note that, in the circuit symbol, the arrow on the emitter lead points in the direction of positive current flow. You can tell whether a transistor in a schematic diagram is PNP or NPN by the direction of the arrow.

The simplest way to think of transistor action is as current amplification: a small current flowing into the base controls a large current flowing into the collector. Both the base and collector currents flow out from the emitter. (This description assumes an NPN transistor. For PNP, the current directions are opposite: a small current flowing *out* from the base controls a large current flowing *out* from the collector, with both currents flowing *in* through the emitter.) The ratio of collector current to base current is called $\beta$ (or $h_{\text{fe}}$) and is typically in the range 20 to 300.

More precisely, however, a transistor is a voltage-controlled current source: small changes in the base voltage cause large changes in collector current. Such a device, in which an input voltage controls an output current, is called a *transconductance amplifier*. The transconductance ($g_m$) for a given device is defined as the change in output current per change in control voltage and has units of $(\text{ohm})^{-1}$ (otherwise known as a *mho*).

To understand the operation of an NPN transistor in more detail, it is convenient to consider the flow of electrons, since electrons are the 'majority carriers' in the N-type regions. Note that the flow of electrons is of course opposite in direction to the flow of conventional positive current. As above,

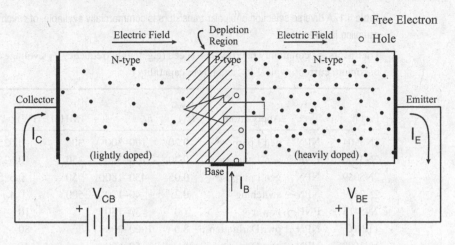

**Fig. 4.2.** Schematic representation of how an NPN transistor operates. External bias voltages create an electric field, which pulls electrons (emitted into the base by the emitter) across the base and into the collector. This results (seemingly paradoxically) in a large flow of electrons through the (reverse-biased) base–collector junction, a current that is easily controlled by small changes in base voltage. The large hollow arrow represents the flow of electrons from the emitter to the collector.

the following description applies also to PNP transistors, but with the current carriers and directions reversed.

In normal transistor operation, the base–emitter diode is forward-biased and the base–collector diode is reverse-biased (Fig. 4.2). The depletion region between the base and the collector extends essentially throughout the thin base region (creating an electric field as shown in Fig. 4.2) and blocks the flow of majority current carriers – holes flowing from base to collector and electrons flowing through the collector to the base. At the same time, the emitter lead injects electrons into the emitter, which flow across the (forward-biased) base–emitter junction. While (as just stated) the base–collector bias inhibits the flow of *holes* from the base into the collector, the *electrons* with which the base is now filled are drawn by the electric field through the junction and into the collector. They do this even though the base–collector junction is reverse-biased. This is the essence of transistor action. Essentially, the construction of the transistor results in large numbers of the 'wrong' current carrier entering the base and then continuing 'downhill' into the collector.

Typically, ≈99% of electrons entering the base from the emitter continue into the collector, and only ≈1% emerge as base current. The base current results from the small fraction of electrons entering the base that combine

**Table 4.1.** A diverse selection of bipolar transistors is commercially available, of which a small sampling is given here.

Transistors are commonly rated by their speed (e.g. toggle frequency $f_T$), voltage capability, maximum current, typical $h_{fe}$, and power capability

| Part # | Type | Application | $I_{C_{max}}$ (A) | $h_{fe}$ | $f_T$ (MHz) | $P_{max}$ (W) | Cost ($US) |
|--------|------|-------------|-------------------|----------|-------------|---------------|------------|
| 2N3904 | NPN | gen'l purpose | 0.20 | 100–300 | 300 | 0.625 | 0.06 |
| 2N3906 | PNP | gen'l purpose | 0.20 | 100–300 | 250 | 0.625 | 0.06 |
| 2N5089 | NPN | gen'l purpose | 0.05 | 450–1800 | 50 | 0.625 | 0.14 |
| 2N2369 | NPN | switching | 0.2 | 40–120 | 500 | 0.4 | 1.40 |
| 2N5415 | PNP | power | 1.0 | 30–150 | | 10 | 0.99 |
| TIP102 | NPN | pwr Darlington | 8.0 | 1000–2000 | | 80 | 0.69 |
| MJ10005 | NPN | pwr Darlington | 20 | 50–600 | | 175 | 6.90 |

with holes. It is desirable to have as large a ratio of collector current to base current as possible. To achieve this, in addition to being as thin as possible, the base is made of lightly doped material.

Since the base–emitter junction obeys the exponential diode law for current as a function of voltage, small changes in the base–emitter voltage difference have a large effect on the collector current. Thus, a transistor can be used as an *amplifier*: a device in which a small signal controls a large signal.

It is worth keeping in mind that $\beta$ is not well controlled in the transistor manufacturing process. Although $\beta$ is approximately constant for a given transistor, it varies from transistor to transistor even if they are of the same type. (For examples of the range of variation, see Table 4.1.) In addition, $\beta$ depends significantly on temperature, collector current, and collector-to-emitter voltage. Since $\beta$ can vary over a substantial range, it is good practice to design transistor circuits in such a way that their proper functioning does not depend strongly on its exact value. Because $\beta$ is so variable, the current-amplifier picture of transistor action is less useful than the transconductance-amplifier picture.

### 4.1.1 Basic definitions

To discuss transistor action quantitatively, we need to define three voltage differences, three currents, and the relationships among them:
- $V_{BE} \equiv V_B - V_E$ = potential of base relative to emitter,
- $V_{CB} \equiv V_C - V_B$ = potential of collector relative to base,

- $V_{CE} \equiv V_C - V_E$ = potential of collector relative to emitter,
- associated identity (from Kirchhoff's voltage law):

$$V_{CE} = V_{BE} + V_{CB}. \tag{4.1}$$

For an NPN transistor in its normal operating mode, all the above potential differences are positive.

- $I_C$ = current flowing into collector,
- $I_E$ = current flowing out of emitter,
- $I_B$ = current flowing into base,
- associated identity (from Kirchhoff's current law):

$$I_E = I_C + I_B. \tag{4.2}$$

For an NPN transistor in its normal operating mode, all the above currents are positive.

The relationships between these voltages and currents are usually expressed in terms of characteristic curves. Fig. 4.3 displays sets of representative curves for an arbitrary bipolar transistor. Study their shapes carefully and refer to them as you perform the following exercises.

## 4.1.2 Simplest way to analyze transistor circuits

In the following section we consider a mathematical model (the *Ebers–Moll* model) that gives a good approximation to transistor performance. However, you don't need the model to understand transistor operation in most circuits. We will see from the Ebers–Moll model that, if a transistor

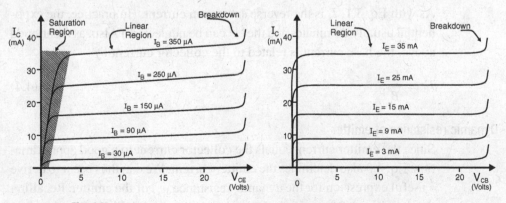

**Fig. 4.3.** Characteristic curves for an NPN bipolar transistor.

is on and conducting milliamperes of collector current, its base–emitter voltage difference $V_{BE}$ is approximately constant at about 700 mV. Furthermore, the base is a point of high impedance, whereas the emitter is a point of low impedance. Thus, a changing voltage at the base causes matching voltage changes at the emitter. In other words, the base voltage controls the emitter voltage:

• the emitter voltage follows the base.

Since $\beta$ is large,

• the collector current nearly equals the emitter current.

Since the collector is of even higher impedance than the base,

• the collector assumes any voltage required by Ohm's law as applied to the rest of the circuit.

These three rules are all you need to know in most situations.

(Suggestion: You may want to skip the next section for now and use these rules to analyze the circuits of Figs. 4.5, 4.7, and 4.8, described in sections 4.2.2, 4.2.3, and 4.2.4. Then come back and study the next section.)

### 4.1.3 Ebers–Moll transistor model

To see why the simple picture just described is valid, we next consider the Ebers–Moll model, which is based on the physical description in section 4.1 and provides a reasonably good description of transistor action. In this model, the amount of collector current that flows is determined by the amount of forward-bias that is applied to the base–emitter diode. We thus have

$$I_C = I_s\left(e^{eV_{BE}/kT} - 1\right). \tag{4.3}$$

As with Eq. 3.1, $I_s$ is the reverse saturation current. (In practice, the exponential usually dominates and the '1' can be neglected.) Also, as mentioned above, the base current is related to the collector current by

$$I_B = \frac{I_C}{\beta}. \tag{4.4}$$

### Dynamic resistance of emitter

Since the emitter current equals the collector current to a good approximation, Eq. 4.3 also detemines the emitter current. We can thus use it to derive a useful expression for the dynamic resistance, $r_e$, of the emitter. Recalling the definition of dynamic resistance, $r_e$ is the partial derivative of emitter voltage with respect to emitter current:

$$r_e \equiv \frac{\partial V_E}{\partial I_E}. \tag{4.5}$$

The dynamic resistance tells us how, for fixed base voltage, the emitter voltage would change in response to a change in emitter current – for example, how the emitter voltage would differ if the emitter were driving a small load resistance as opposed to a large one. To determine $r_e$, we differentiate Eq. 4.3 at fixed base voltage, giving

$$r_e = \frac{kT}{e} \frac{1}{I_C}. \tag{4.6}$$

(In addition there is the ohmic resistance of the emitter, which is typically a few ohms.)

### Dynamic resistance of base

On the other hand, if we fix the emitter voltage, we can find from Eqs. 4.3 and 4.4 the dynamic resistance $r_{BE}$ to the emitter as seen at the base:

$$r_{BE} = \frac{\partial V_B}{\partial I_B} = \beta \frac{kT}{e} \frac{1}{I_C}. \tag{4.7}$$

This tells us how the base loads the circuit that is driving it. We see that the base–emitter junction appears to have a low resistance when viewed from the emitter end, but appears to have a higher resistance (by a factor $\beta$) when viewed from the base end.

### Some useful approximations

Since at room temperature, $e/kT = 39$ V$^{-1}$, in practice a reasonable approximation is

$$r_e = \frac{25 \text{ mV}}{I_C}, \tag{4.8}$$

$$r_{BE} = \beta \frac{25 \text{ mV}}{I_C}. \tag{4.9}$$

Since the emitter acts as a low impedance (only a few ohms for typical collector-current values), its voltage hardly depends on the current flowing through it. But the base acts as a high impedance, so it is easy to apply a signal voltage to the base. This comes about because the base current is smaller than the emitter current by the factor $\beta$, which is of order 100. Although $r_{BE}$ is only a few hundred ohms, the factor $\beta$ applies also to any resistor

$R_E$ that is in series with the emitter, i.e. the apparent resistance seen at the base is $\beta(r_e + R_E) = r_{BE} + \beta R_E$, which is typically tens to hundreds of kilohms.

We saw in the case of the silicon diode that a crude approximation in which the forward diode drop is taken to be approximately constant at 600 mV is adequate for most applications. As mentioned above, in many practical transistor applications (including the circuits you will build in this chapter) a simple approximation is sufficient: treat the base–emitter voltage difference $V_{BE}$ as constant at about 700 mV and $r_e$ as constant at a few ohms. This reflects the fact that the order of magnitude for $I_C$ in a typical small-signal-transistor circuit is several milliamperes. Often $r_e$ is much smaller than $R_E$ and can be neglected.

## 4.2 Experiments

### 4.2.1 Checking transistors with a meter

Since a transistor is constructed as a pair of back-to-back PN junctions, a quick way to test a transistor is to verify its junction resistances in the forward- and reverse-biased directions. Often this can be done using an ohmmeter, which sends current through the device under test and measures the resulting voltage. However, our digital multimeters are not designed for this kind of measurement; instead, they provide a *diode test* function that measures the forward voltage corresponding to a forward current of about 600 μA. To test a diode or transistor junction, connect it between the 'VΩ' and 'common' jacks with clip leads and set the meter's selector knob to the position marked with the diode symbol. The diode is forward-biased if its anode is connected to 'VΩ' and its cathode to 'common'.

▷ Test the base–collector and base–emitter junctions of a 2N3904 (NPN) and 2N3906 (PNP), and record your readings in both the forward- and reverse-biased directions.

To tell which pin of the transistor is which, refer to Fig. 4.4, which shows the pinout for the TO-92 plastic case in which these transistors are packaged. If your transistors are good, each junction should show about 700 mV in the forward direction, and an out-of-range indication in the reverse direction.

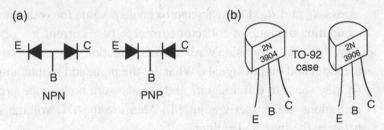

**Fig. 4.4.** Transistor as back-to-back diodes; TO-92 pinout.

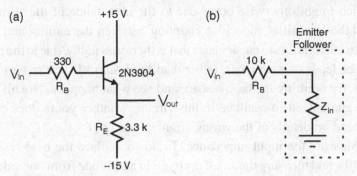

**Fig. 4.5.** (a) Emitter follower. (b) Emitter-follower model used for input-impedance measurements. The value for $Z_{in}$ is found using the voltage-divider equation.

## 4.2.2 Emitter follower

This simple transistor circuit (shown in Fig. 4.5(a)) is often used to 'buffer' an AC signal, as well as to change its DC voltage level by $V_{BE}$. (It is also a close relative of the common-emitter amplifier that we will study next.) It is called an emitter follower because the voltage at the emitter follows voltage changes at the base, with $V_{BE}$ almost constant at about 0.7 V. Although it has voltage gain ($V_{out}/V_{in}$) of about unity (actually slightly less because of the logarithmic dependence of $V_{BE}$ on $I_C$ implied by Eq. 4.3), it is still an amplifier: it has high input impedance, and low output impedance, and can thus provide current gain (i.e., output current > input current), which is what is meant by buffering.[1]

▷ Start by grounding $V_{in}$. You can determine $I_B$ and $I_C$ (within the uncertainties due to the resistor tolerances) by measuring the voltage drops

---

[1] Since power is the product of voltage and current, power amplification can occur through an increase in either quantity.

across $R_B$ and $R_E$. Derive an approximate $\beta$ value for your transistor by computing the ratio of collector current to base current.

▷ Now drive $V_{in}$ with a sine wave from the function generator and compare the input and output signals. What are the input and output amplitudes? (If they seem to differ significantly, make sure both scope-probe compensations are properly adjusted.) Measure the DC voltage shift $V_{BE}$ between the base and emitter.

The 330 $\Omega$ base resistor is in series with the input resistance $r_{BE} + \beta R_E$, so it has little effect on the signal. It is there to prevent parasitic oscillation, which might otherwise occur due to the inductance of the wire jumpers and the 'parasitic' capacitive coupling between the emitter and the base. ('Parasitic' refers to capacitance that is there inevitably, due to the proximity of the leads to each other, rather than by design.) If you're curious about this, try omitting the base resistor and see what happens. Not all 2N3904s are guaranteed to oscillate in this circuit; whether yours does could also depend on details of the wiring arrangement.

▷ Measure the input impedance. To do so, replace the base resistor with 10 k and measure the small decrease in amplitude from one side of $R_B$ to the other (Fig. 4.5(b)). Explain using the voltage-divider idea how this measures the input impedance. (When you're done with this exercise, restore the 330 $\Omega$ base resistor for use in the following parts.)

▷ Measure the output impedance. To do so, add a blocking capacitor and load resistor as shown in Fig. 4.6(a) and infer the output impedance from the small decrease in output amplitude (see Fig. 4.6(b)).

The blocking capacitor allows the 330 $\Omega$ load to affect the AC signal voltage without changing the DC biasing of the transistor.

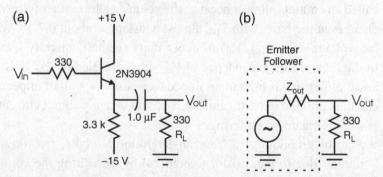

**Fig. 4.6.** (a) Emitter follower with optional load circuit for measurement of $Z_{out}$. (b) Emitter follower modeled as an ideal voltage source in series with an output impedance.

▷ Explain what would happen to the DC bias voltage of the emitter if the capacitor were omitted.

When using the blocking capacitor, be sure to use a small signal amplitude (about 1 V) so as not to apply too large a reverse voltage to the capacitor, and use a high enough frequency so that the capacitor causes negligible attenuation – about 10 kHz. Note that polarized capacitors can be safely reverse-voltaged by a volt or two r.m.s., but typically not by more than 15% of their voltage rating.

▷ How do your measured impedances compare with what you expect?

The input impedance should equal $\beta$ times the emitter resistor, and the output impedance should equal the dynamic resistance of the emitter (as described in section 4.1.3).

### 4.2.3 Common-emitter amplifier

The *common-emitter* amplifier is a very common transistor amplifier configuration, but that is not how it gets its name – the name reflects the idea that the emitter is in common between the input circuit and the output circuit. Construct the common-emitter amplifier shown in Fig. 4.7.

▷ Predict and measure the *quiescent* DC voltages (i.e. the voltages when no input signal is present) at the base, emitter, and collector.

The predictions are easy:

1. Apply Ohm's law to the base-bias resistive voltage divider to determine the quiescent base voltage: $V_B = V_{CC} R_2 / (R_1 + R_2)$. This is an approximation since it neglects the base current, but, as you'll see, given the large value of $\beta$, the base current is small enough that the approximation is a good one.

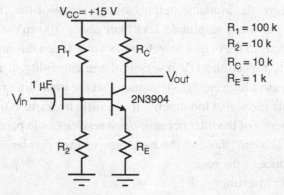

$$V_{CC} = +15\ V$$

$R_1 = 100\ k$
$R_2 = 10\ k$
$R_C = 10\ k$
$R_E = 1\ k$

**Fig. 4.7.** Common-emitter amplifier.

2. Then apply Ohm's law to the emitter resistor to determine the emitter current, taking into account the transistor's expected $V_{BE}$ drop: $I_E = (V_B - V_{BE})/R_E$.

3. Then apply Ohm's law to the collector resistor to determine the quiescent collector voltage: $V_{out} = V_{CC} - I_C R_C$. You know the collector current well enough since it equals the emitter current to a good approximation.

▷ Compared to your measurements, by what percentages are your voltage predictions wrong? Is this as expected given the resistor tolerances and uncertainties in $\beta$ and $V_{BE}$?

▷ Using the measured voltages, predict the collector, emitter, and base currents.

▷ Calculate the change in quiescent base voltage if you take the base current into account. Assume that the base current flows through the Thévenin equivalent of the base-bias voltage divider (i.e., the input impedance of the base is in parallel with $R_2$).

You can understand how the circuit amplifies by applying Ohm's law to the emitter and collector resistors. Since the emitter follows the base, a voltage change at the base causes a larger voltage change at the collector:

$$\Delta V_{out} = -\Delta I_C \cdot R_C = -\Delta I_E \cdot R_C$$

$$\Delta I_E = \Delta V_E/R_E = \Delta V_B/R_E = \Delta V_{in}/R_E, \tag{4.10}$$

Therefore,

$$\Delta V_{out} = -\Delta V_{in} R_C/R_E. \tag{4.11}$$

▷ Measure the voltage gain ($\Delta V_{out}/\Delta V_{in}$) and compare with what you expect.

So as not to exceed the available output-voltage range of the circuit, be careful to keep the input amplitude less than about 700 mV. (You can check what happens to the output waveform as you exceed this amplitude, but be sure not to exceed the 1 V reverse-voltage capability of the input capacitor.) Also, use a high enough frequency that the input high-pass filter does not attenuate the signal too much – it is a little tricky to estimate the breakpoint frequency of the filter because *three* resistances in parallel need to be taken into account: those of the base-bias voltage divider as well as the input impedance of the base.

▷ Is the amplifier inverting?

▷ Look at the signal at the emitter and explain what you see.

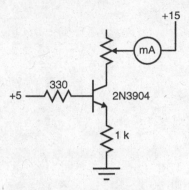

**Fig. 4.8.** Transistor current source.

▷ Try a triangle-wave input – can you see any distortion in the output waveform? There should be some due to the variation of $r_e$ with collector current, but the effect is small since $r_e$ is in series with the 1 k emitter resistor. How big should the effect be according to the Ebers–Moll model?

In the *grounded-emitter* amplifier, i.e., for $R_E = 0$, the voltage gain is greater, but so is the distortion, since $r_e$ alone appears between the emitter and ground.

### 4.2.4 Collector as current source

Since the base–collector junction is reverse-biased, the collector should act as a very high impedance. This means that the collector is a good approximation to a current source: a device that outputs a constant current, independent of its voltage. You can verify this using the circuit of Fig. 4.8, using a second meter or the scope to make various voltage measurements. Start with the load resistance (10 k pot – as always, be careful to connect it properly so as not to burn it out!) set to 0 $\Omega$.

▷ How much collector current should flow? Is this confirmed by your measurement of the emitter voltage?

▷ Slowly increase the load resistance until the output current starts to decrease. At the pot setting where the current source starts to fail (output current starts to vary rapidly with load resistance), measure the collector voltage. What is the *compliance* of your current source (the range of output voltage over which the current is approximately constant)?

▷ Where the current source starts to fail, how does the collector voltage compare with the base voltage? You should be observing transistor *saturation*: when $|V_{CE}| < |V_{BE}|$, the current-source behavior stops since

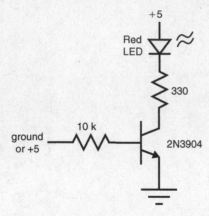

**Fig. 4.9.** Transistor switch.

the collector–base junction becomes forward-biased and its impedance decreases.

Another way of thinking about the same phenomenon is that when the transistor saturates, $\beta$ decreases sharply, thus the base current increases as the base 'steals' current from the collector.

▷ By measuring the voltage drop across the base resistor, take a few measurements of $\beta$ as you turn up the load resistance and the transistor becomes more and more saturated. Make a graph of $\beta$ vs. $V_{CE}$.

### 4.2.5 Transistor switch

Transistor saturation is put to good use in the *saturated switch*. Construct the circuit of Fig. 4.9. You should see the light-emitting diode (LED) turn on when you connect the input to +5 V and off when you leave it open or connect it to ground. (If you are unsure which terminal of the LED is the anode and which is the cathode, you can check it with a meter, or feel free to try it both ways. Often the cathode is indicated by a flat spot on the rim of the plastic that encapsulates the diode, and some manufacturers make the anode lead slightly longer than the cathode lead.)

Note that the transistor in this common-emitter connection is an 'inverter', in the sense that a high voltage level at its input (the open end of the series base resistor) causes a low voltage level at its output (the collector). This circuit is called a saturated switch since the transistor goes into saturation ($|V_{CE}| < |V_{BE}|$) when turned on. It has the virtue of dissipating very little power, since, when the transistor is on, the voltage across it is small, whereas, when the transistor is off, the current through it is essentially zero.

▷ When the transistor is on, what are the base and collector currents? (You don't need to bother with an ammeter for this measurement – just look at the voltage drops across the base and collector resistors.)

▷ What is your transistor's saturation voltage $V_{CE(sat)}$? What is the 'on voltage' across the LED?

▷ Approximately what minimum value of $\beta$ must the transistor have to be sure of saturating when $+5$ V is applied at the input?

▷ Drive the switch from the 'TTL' output of the function generator ('digital' square-wave with a low voltage level near zero and a high level near $+5$ V) at 100 kHz and use the dual-trace oscilloscope to measure the turn-on and turn-off delays in nanoseconds. (Trigger the scope with the function generator while looking at both the function generator signal and the collector voltage.)

The relatively slow turn-on and turn-off delays of the saturated switch are due to the charge stored in the base when the transistor saturates. It takes time for the transistor to switch states since this saturation charge must flow in or out of the base through the input resistor. High-speed switching transistors (such as the 2N2369) are manufactured to minimize this effect and can operate at frequencies as high as 1400 MHz.

## 4.3 Additional exercises

The following optional exercises offer additional practice.

### 4.3.1 Darlington connection

To provide high input impedance and reduce the input base current, one can cascade two transistors in series, i.e. 'buffer' the input with an emitter-follower stage. This *Darlington* transistor pair acts like a single transistor whose current gain is the product of the two $\beta$'s and whose $V_{BE}$ drop is the sum of the two $V_{BE}$'s. Build the circuit of Fig. 4.10.

▷ With the input grounded, what quiescent currents do you observe through $R_B$ and $R_E$? What does this imply for the combined $\beta$ value of the Darlington pair?

▷ Now apply an input signal – what do you see at the output? What is the DC voltage drop from input to output?

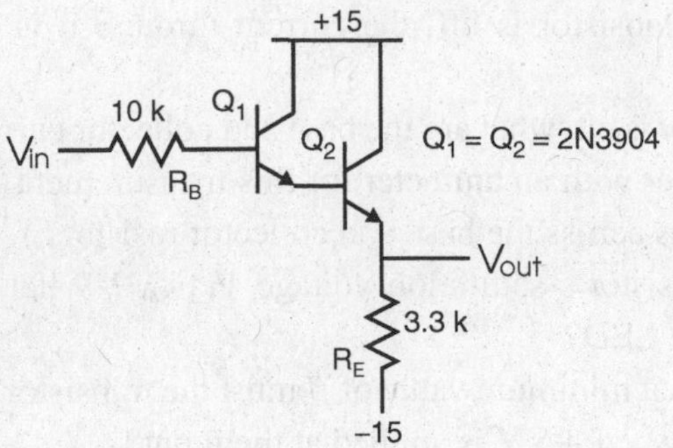

**Fig. 4.10.** Darlington pair.

▷ The input impedance should be so big that you can't measure any decrease in signal amplitude across the 10 k resistor – check this assumption. What minimum value does this imply for the input impedance? What input impedance do you expect, and why?

Darlington pairs are available encapsulated in three-lead packages, for example the 2N6426 with combined $\beta$ value of about 100 000. The Darlington connection is particularly useful for power transistors, to compensate for their low $\beta$ ($\beta \approx 20$ is not uncommon). For example, the TIP110 50 W power Darlington has a minimum combined $\beta$ value of 500.

### 4.3.2 Push–pull driver

To provide low output impedance, a push–pull buffer stage is often used. This consists of two emitter followers, one PNP, and one NPN, arranged so that the PNP conducts during one half of the output period (when $V_{out} < 0$) and the NPN during during the other half ($V_{out} > 0$).

▷ First, drive your breadboard's speaker (nominal impedance = 8 Ω) from the function generator. With the amplitude set to maximum and the frequency at 1 kHz, measure the function-generator amplitude both with and without the speaker connected. What is the attenuation due to the 8 Ω load? Is it consistent with the measurement of the function generator's output impedance you made in section 3.7?

▷ Since the speaker impedance may depend on frequency, repeat the measurement at 10 kHz and compare.

▷ Go back to 1 kHz and add a push–pull buffer as shown in Fig. 4.11. (Be sure you have adjusted the supply voltages to not more than ±5 V so as not to overheat the transistors – they are rated for 200 mA collector current and 625 mW power dissipation.) You should see a larger output amplitude and hear a louder tone from the speaker.

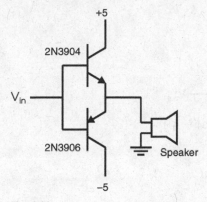

**Fig. 4.11.** Driving loudspeaker with push–pull buffer.

The circuit may oscillate because of inadvertent positive feedback from output to input; if it does, try (1) rearranging your circuit so that the wire jumpers are as short as possible, (2) adding a 330 Ω resistor in series with the input, and if that's not enough to stabilize it, (3) adding a few-hundred-picofarad cap to ground at the output.

▷ The output waveform will display 'crossover distortion' – what does it look like and why does it occur? (Hint: for a transistor to be on, there must be about 700 mV between base and emitter – is there a time during the cycle when neither transistor is on?) By how much does it reduce the output amplitude, and why? What is the minimum input amplitude required for an audible output? Explain.

For high-power applications, a power-Darlington push–pull stage is often used. Two methods can be used to alleviate the crossover distortion:

• shifting the bias points of the two transistors apart to minimize the portion of the cycle when neither transistor is on;

• using feedback to apply a signal to the bases that compensates for the distortion.

We will explore these techniques in Chapter 8.

### 4.3.3 Common-base amplifier

Build the common-base amplifier of Fig. 4.12.

▷ Predict and measure the transistor's quiescent currents and bias voltages. The diode at the base should bias the emitter approximately at ground. Since the diode and base–emitter voltage drops are unlikely to be exactly the same, the input will have a small DC offset – how big is it?

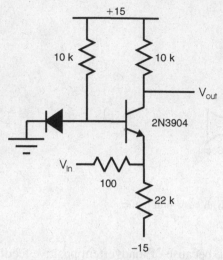

**Fig. 4.12.** Common-base amplifier.

▷ Connect a small sine-wave input and determine the voltage gain. Note that, in contrast to the common-emitter amplifier, the input and output currents are almost equal, and the amplifier is noninverting.

▷ What do you predict for the input and output impedances? You can measure the input impedance easily using the 400 Ω output impedance of the function generator: how much smaller does the function-generator output become when you connect it to the amplifier input? What input impedance does this imply for the amplifier? Explain.

# 5 Transistors II: FETs

In this chapter we introduce the field-effect transistor (FET). A majority of today's integrated circuits are built using FETs of one type or another. FET operation is easier to explain than that of bipolar transistors; however, due to the variability of FET parameters, many people find FETs more difficult to use. As with bipolar technologies, it is essential that you master the basics of FET operation, and you will find that knowledge useful later on.

## Apparatus required

Breadboard, oscilloscope, multimeter, two 2N5485 JFETs, one 1N4733 Zener diode, two 1 k, one 3.3 k, two 10 k, one 100 k, and one 1 M $\frac{1}{4}$ W resistors, 0.1 μF ceramic capacitor, 1.0 μF and 100 μF electrolytic capacitors.

## 5.1 Field-effect transistors

Like bipolar junction transistors, field-effect transistors (FETs) are three-terminal semiconductor devices capable of power gain. Qualitatively, they operate much like junction transistors, but they have much higher input impedance and lower transconductance and voltage gain. Also, they have a larger variation in their '$V_{BE}$' equivalent (called $V_{GS}$) than bipolar transistors. They come in a confusing variety of types, but we will concentrate for today on junction FETs (JFETs).

Fundamentally, there are two types of FETs: junction FETs and metal-oxide-semiconductor FETs (MOSFETs). In both types, a conducting *channel* between the *drain* and *source* terminals is controlled by a voltage applied to the *gate* terminal. The channel can be made of either N-type or P-type material (Fig. 5.1). N-channel is more common since the conductivity of N-type semiconductor (in which electrons carry the current) is higher than that of P-type (in which holes do).

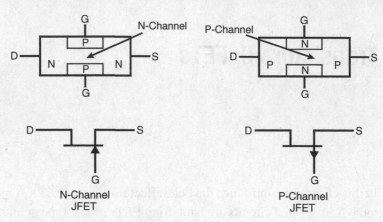

**Fig. 5.1.** Construction and circuit symbols of JFETs (note that other variants of these symbols are also used).

The gate region of a JFET consists of material of opposite type to that of the channel; thus, the gate and channel form a diode. In our preferred symbol for JFETs (Fig. 5.1), to distinguish the source from the drain, the gate terminal is drawn at the source end, even though the channel is physically spread out between the source and drain. As for any diode, the arrow on the gate symbol indicates the direction of forward-bias. However, JFETs are normally used with the gate–channel diode reverse-biased.

In a JFET, the channel conducts unless it is turned off by an applied reverse-bias voltage between the gate and the channel. As the reverse-bias is increased, more and more current carriers are repelled out of the channel until it is 'pinched off' and the drain–source current drops to zero (see Fig. 5.2). The voltage at which this occurs is called $V_P$ or $V_{GS(off)}$.

Note that the drain, gate and source of an FET play similar roles to the collector, gate, and emitter (respectively) of a bipolar transistor. Unlike the bipolar case, the source and drain are roughly interchangeable and it is possible for an FET to be used 'backwards'.

### 5.1.1 FET characteristics

The simplest way to think of FET action is as a voltage-controlled current source, i.e. the drain current $I_D$ is approximately constant for a given gate–source voltage $V_{GS}$, depending only slightly on the voltage $V_{DS}$ between the drain and source. Since the gate–channel diode is normally reverse-biased, the gate current is extremely tiny (typically ~ nanoamperes), so that for all

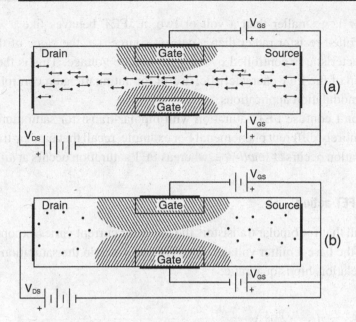

**Fig. 5.2.** Schematic representation of JFET operation: (a) gate–channel diode slightly reverse-biased; (b) gate–channel diode highly reverse-biased ($V_{GS} \geq V_P$ so that channel is pinched off).

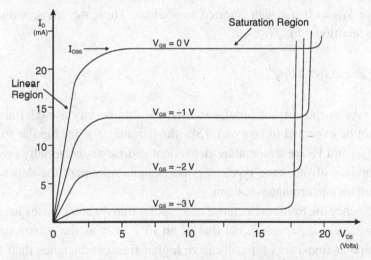

**Fig. 5.3.** Idealized common-source characteristic curves for a JFET.

practical purposes the drain and source currents are equal ($I_D = I_S$). The voltage-controlled current-source behavior occurs as long as the drain–source voltage $V_{DS}$ is sufficiently high. This is called the *saturation* region of the FET characteristic (see Fig. 5.3).

For $V_{DS}$ smaller than a volt or two, a JFET behaves like a voltage-controlled resistor rather than a current source, i.e. the *slope* of the $I$–$V$ characteristic is controlled by the gate–source voltage. This is the *linear* region of the FET characteristic, useful for automatic gain control (AGC) and modulation applications.

Don't confuse FET saturation with bipolar-transistor saturation – they are entirely different phenomena! For example, recall that bipolar-transistor saturation occurs at *small* $V_{CE}$, whereas FET saturation occurs at *large* $V_{DS}$.

### 5.1.2 Modeling FET action

Recall that for bipolar transistors the collector current varies exponentially with the base–emitter voltage. For FETs operated in the saturation region, the relationship is quadratic:

$$I_D = I_{DSS} \left( 1 - \frac{V_{GS}}{V_P} \right)^2 , \tag{5.1}$$

where $V_P$ is the pinch-off voltage and $I_{DSS}$ is the saturation drain current for $V_{GS} = 0$ (i.e. gate shorted to source). Thus, the transconductance is proportional to $\sqrt{I_D}$:

$$g_m = \Delta I_D / \Delta V_{GS} . \tag{5.2}$$

As in the case of bipolar transistors, this is only a model and should not be expected to be exact. Like the parameter $\beta$ for bipolar transistors, $I_{DSS}$ and $V_P$ are temperature-dependent and vary substantially even among devices of the same type, so good designs minimize the dependence of circuit performance on them.

Since the transconductance of a bipolar transistor increases linearly with $I_C$ $((g_m)_{\text{bipolar}} = 1/r_e)$, but that of an FET only as the square root of $I_D$, bipolar transistors typically have higher transconductance than FETs for a given current, and thus can give higher gain in amplification applications. This has led to the common practice of combining FETs with bipolar transistors in analog integrated circuits to exploit the advantages of both, e.g. the CA3140 MOSFET-input op amp with its teraohm input impedance.

## 5.2 Exercises

### 5.2.1 FET characteristics

As shown in Fig. 5.1, the gate–source and gate–drain connections are PN junctions.

▷ Use the diode-test feature of the multimeter and verify this picture using a 2N5485 JFET (the pinout for the 2N5485 is shown in Fig. 5.4). **Note:** the pinout of the 2N5485 does **not** correspond to that of the 2N3904.

Unlike an NPN transistor, whose emitter and collector are distinct N-type regions separated by the P-type base, the drain and source of an N-channel JFET occupy opposite ends of a single N-type region, connected via the channel.

▷ Use an ohmmeter to show that the drain and source are connected. If the meter reading fluctuates, try connecting the gate and source together using the breadboard and a small piece of wire. Explain why this will stabilize your measurement. What resistance do you measure?

Next verify Eq. 5.1. To measure $I_{DSS}$ and $V_P$, set the drain voltage to 10 V, ground the source, and apply a variable negative voltage to the gate. This can be accomplished using either the 1 k or 10 k potentiometer as a voltage divider (ground one end of the pot, set the other end to −5 V, and connect the slider to the gate).

▷ Measure $I_{DSS}$. Adjust the pot until the gate is at ground while leaving the drain voltage at 10 V. According to Eq. 5.1, the drain current now equals (by definition) $I_{DSS}$.

▷ Measure $V_P$. Using an ammeter to measure the drain current, adjust the gate voltage until the drain current drops to zero. The pinch-off voltage will be equal to this gate voltage. Why is this true?

▷ Verify Eq. 5.1. With the drain voltage set to 10 V, adjust $V_{GS}$ while measuring the drain current. Plot $I_D$ versus $V_{GS}$.

The common-source characteristic curves for the 2N5485 can be measured using the circuit shown in Fig. 5.4. These curves illustrate the basic dependences among $I_D$, $V_{DS}$, and $V_{GS}$, and will be useful while performing the remaining exercises.

Since JFETs are notoriously variable, more so than bipolar transistors, try to use the same JFET for all exercises. Use one pot to adjust the gate

(a)                    (b)

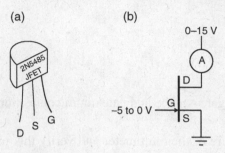

**Fig. 5.4.** Circuit for measuring the common-source characteristic curves.

voltage and the other to adjust the drain voltage. Use the scope probes to measure the drain and gate voltages while using a meter to measure the drain current.

▷ Using the pot, adjust $V_{GS}$ to be 0.5 V more positive than $V_P$ (keep in mind that $V_P$ is negative!). Slowly increase the drain voltage from zero to 15 V while measuring the drain current. Record and plot your measurements.

For $V_{DS}$ less than a few volts, the current should increase linearly with drain voltage. This is the 'linear region', in which the JFET acts as a voltage-controlled variable resistor. As you further increase $V_{DS}$, the current should then 'saturate' at an approximately constant value.

JFET saturation occurs because the increasing drain voltage creates an increasing depletion region between the gate and drain. Since $V_{GS} > V_P$, the channel will never be pinched off completely, with an equilibrium (of sorts) created. The size of the depletion region (and thus the resistance of the channel) increases approximately linearly with drain–source voltage difference, resulting in approximately constant current.

▷ Repeat your measurement procedure for several $V_{GS}$ values between $V_P$ and zero. Plot the data on a single graph and clearly label each curve, indicating the linear region, the saturation region, and $I_{DSS}$.

## 5.2.2 FET current source

Adding a resistor improves the JFET's current-source performance compared with that of the bare JFET.

▷ Hook up the circuit shown in Fig. 5.5 and measure the dependence of the drain current on the drain–source voltage as you adjust the pot; record and plot your measurements.

▷ What is $V_{GS}$ for each data point?

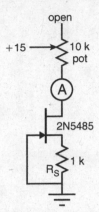

**Fig. 5.5.** Self-biasing JFET current source.

You can compute $V_{GS}$ from $I_D$ and the known resistance of $R_S$. You should see the drain current start to vary substantially as you make the transition from the saturation region to the linear region.

▷ What is the compliance?

▷ Within the saturation region, how constant is the current?

▷ Calculate the approximate output impedance in the saturation region (see Eq. 3.13).

▷ Compare the performance of this current source with that of the bare JFET and of the bipolar current source that you built in section 4.2.4.

Even though $V_{GS}$ is not exactly constant as $V_{DS}$ is varied, this circuit actually works *better* (has larger output impedance) than one in which $V_{GS}$ is held constant. This is because *negative feedback* is at work. For example, suppose $I_D$ increases; then, so does the drop across $R_S$, increasing the magnitude of $V_{GS}$ and moving the FET closer to pinch-off, thus decreasing $I_D$.

### 5.2.3 Source follower

As with the emitter follower from section 4.2.2, a JFET source follower provides power amplification via current buffering, even though the voltage gain is less than or equal to unity. Build the circuit of Fig. 5.6, and try it out with a 1 kHz sine wave of small amplitude.

The operation of this circuit is reasonably straightforward. Given the nanoampere gate current, the 1 M resistor is small enough to bias the gate very near ground. Quiescently, the FET is thus in the saturation region, with $I_D$ determined by $I_{DSS}$ and $V_P$ according to Eq. 5.1. If the input voltage increases, $V_{GS}$ moves closer to zero, the channel opens, and $I_D$ increases.

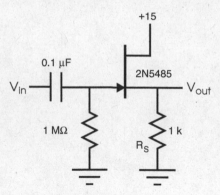

**Fig. 5.6.** Source follower.

Thus, $V_{\text{out}}$ follows the source. When $V_{\text{in}}$ decreases, the channel closes and $V_{\text{out}}$ drops.

▷ What is the DC offset at the output?

▷ Measure the voltage gain.

You should see that the voltage gain is less than unity, since the dynamic resistance of the source ($=1/g_m$) forms a voltage divider with $R_S$. (This effect was also present for the bipolar transistor, but was much smaller due to the bipolar transistor's larger value of $g_m$.) Draw a diagram of this voltage divider.

▷ From your observed attenuation, derive a value for $g_m$ and compare with that of a bipolar transistor at the same current.

You can improve the source follower by providing it with much higher load resistance. Since an ideal current source would have infinite resistance, a current-source load is often used; it can be constructed by adding another FET, as in the clever circuit of Fig. 5.7. (Since we are using N-channel JFETs, it is actually a current sink.) Try it out.

▷ Measure the voltage gain and the DC offset from input to output.

Note that if the two 2N5485s approximately match in their characteristics, not only is the voltage gain unity, but the DC offset is small: the constant current due to $Q_2$ creates a constant voltage drop across $R_1$. Furthermore, since the gate of $Q_2$ is at the same voltage as the bottom of $R_2$, to the extent that the two FETs (and the two resistors) match, this should also be true for $Q_1$ and $R_1$. Thus the output voltage must follow the source voltage of $Q_1$.

▷ Explain the operation of this circuit in your own words.

▷ What are $I_D$, $V_{\text{GS1}}$, and $V_{\text{GS2}}$?

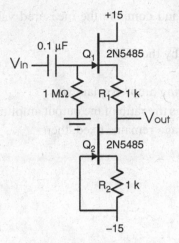

**Fig. 5.7.** Source follower with current-source load.

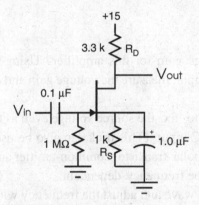

**Fig. 5.8.** JFET amplifier.

The offset can be much improved by using a matched FET pair, e.g. the 2N3958 dual JFET. Such a circuit is often used in the input stage of an oscilloscope.

## 5.2.4 JFET amplifier

Construct the amplifier shown in Fig. 5.8. The principle at work is essentially the same as for the common-emitter amplifier: a varying input voltage controls a varying current, which the drain resistor 'converts' to an output voltage.

▷ Using the common-source curves that you measured in section 5.2.1, predict the quiescent gate–source bias voltage and output voltage for this amplifier. Power the amplifier and compare the measured values with your predictions.

▷ How much power is dissipated by the FET?

▷ What is the input impedance?

▷ Is this an inverting or noninverting amp? Explain why.

The voltage gain ($A$) is defined as the ratio of the output amplitude to the input amplitude. If the source voltage remains fixed, then

$$\Delta V_{\mathrm{G}} = \Delta V_{\mathrm{GS}}.$$

As discussed previously,

$$g_m = \Delta I_{\mathrm{D}} / \Delta V_{\mathrm{GS}},$$

and since

$$\Delta V_{\mathrm{D}} = \Delta I_{\mathrm{D}} \cdot R_{\mathrm{D}},$$

$$A = g_m \cdot R_{\mathrm{D}}.$$

▷ What is the predicted voltage gain for this amplifier? Using a 1 kHz small-amplitude sine-wave input, measure the voltage gain and compare with the expected gain.

The source capacitor is used to 'fix' the source voltage even as the drain current fluctuates due to the AC input. (This trick can also be used to increase the voltage gain of the bipolar-transistor common-emitter amplifier.) This implies that the gain will be frequency-dependent.

▷ Switch the input to a triangle wave and adjust the frequency widely. Explain what you see. Replace the source capacitor with a 100 μF capacitor. How does this change things?

▷ Try several different 2N5485s and record the voltage gain and quiescent drain current and output voltage. How reproducible are the results?

▷ Comment on the design and operation of simple transistor and JFET circuits. For example, when would you choose a bipolar transistor over a JFET or vice versa? Feel free to include any general comments you have on the experience you've gained from the last few chapters.

# 6 Transistors III: differential amplifier

In this chapter we will study the transistor differential-amplifier circuit. This is a very important transistor circuit, as it is the basis of the operational amplifier (or *op amp*), one of the most useful devices for analog signal processing. Probably the most surprising thing about op amps is their very large voltage gain, usually exceeding 100 000. This chapter will give you a clearer idea how such performance is achieved. We will also look at some other circuits that serve as building blocks for op amps.

## Apparatus required

Breadboard, oscilloscope, multimeter, three 2N3904 and three 2N3906 transistors, one 5.1 V Zener diode, three 100 Ω, five 10 k, two 22 k one each of 560 Ω, 2.2 k, and 100 k $\frac{1}{4}$ W resistors.

## 6.1 Differential amplifier

A *differential amplifier* is an amplifier for differential input signals, i.e. it amplifies the voltage *difference* of its two inputs. This is useful in two important ways:

1. A differential amplifier can be used to amplify a differential signal (the voltage difference between the two inputs) while suppressing any noise that is common to the two inputs.
2. As we will see in future chapters, differential amplifiers make it easy to build circuits that use negative feedback.

Don't confuse the differential amplifier with the *differentiator*: although the names sound similar, the two circuits perform entirely different operations.

▷ What does each do?

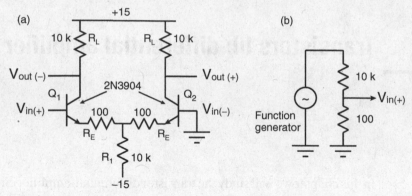

**Fig. 6.1.** (a) Differential amplifier; (b) function generator with 100-to-1 attenuator.

### 6.1.1 Operating principle

Fig. 6.1(a) shows a differential amplifier made of two NPN transistors. You can think of it as a 'current divider'. Quiescently, the current through $R_1$ is shared equally between $Q_1$ and $Q_2$. If a differential input signal is applied to the bases of $Q_1$ and $Q_2$, some collector current shifts from one transistor to the other. This change, $\Delta I$, causes a positive voltage change at one collector and a negative voltage change at the other, i.e. a differential output signal, as explained in more detail below.

### 6.1.2 Expected differential gain

The differential voltage gain can be understood as follows. Suppose we apply a differential signal consisting of equal and opposite changes in the base voltages of $Q_1$ and $Q_2$:

$$\Delta V_{\text{in}+} = -\Delta V_{\text{in}-} \equiv \Delta V. \tag{6.1}$$

Since the emitter voltages follow the base voltages, a similar voltage change occurs between the two emitters. This causes a current $\Delta I$ to flow across the two emitter resistors. Since these are in series with the dynamic emitter resistances of the two transistors (each of which has an approximate value $r_e = 1/g_m = 25\,\text{mV}/I_C$ according to the Ebers–Moll model), we have

$$\Delta I = \Delta V/(R_E + r_e). \tag{6.2}$$

Since a purely differential signal raises the voltage at one emitter by the same amount that it lowers the voltage at the other emitter, it does not change the voltage at the top of $R_1$; thus, the current through $R_1$ is constant. This means that the current due to the differential signal must add to the collector current of one transistor and subtract from that of the other transistor, so the differential voltage gain should be $R_L/(R_E + r_e)$. One-half of the amplified differential signal appears at each collector. In other words, a small differential input voltage applied to the two bases causes a large differential output signal at the two collectors.

### 6.1.3 Measuring the differential gain

Wire up the circuit, check it against your schematic (Fig. 6.1), and try it out. Because it has a large gain, you will need to use a small enough amplitude out of the function generator so that the amplified signal is not 'clipped' (cut off at the top or bottom). However, when the function generator is set to a small amplitude, it puts out a rather noisy signal with a relatively large DC offset. To avoid these problems, run the function-generator output through a 100-to-1 attenuator (voltage divider) made from a 10 k resistor in series with a 100 $\Omega$ resistor to ground (Fig. 6.1(b)).

▷ Measure the attenuation (voltage-division ratio) by setting the function generator for a large amplitude and measuring the signal amplitudes before and after the voltage divider. Compare your observed attenuation with the theoretical value.

Now that you know the attenuation, you can display the function-generator output on the scope and calculate from it the size of the actual input signal to the amplifier.

Connect the output of your attenuator to the base of $Q_1$, and observe the amplifier outputs at the collectors of $Q_1$ and $Q_2$.

▷ Measure the differential voltage gain $A_{diff} = \Delta(V_{out+} - V_{out-})/\Delta(V_{in+} - V_{in-})$ for a few different input amplitudes and frequencies; compare with what you expect.

The amplifier circuit clips its output when all of the available current has been switched to one transistor or the other; this determines maximum and minimum voltages, beyond which the output cannot go.

▷ Try it and see. At what output voltages does clipping set in? Compare with what you expect – you can estimate the amount of current that is available to either transistor by measuring the voltage drop across $R_1$.

### 6.1.4 Input offset voltage

If you ground both inputs, what are the output voltages? For an ideal differential amplifier, they should be equal, but you will probably find that, due to small mismatches between the two collector resistors, the two emitter resistors, and the two transistors, they are not. To obtain exactly equal outputs you would have to input a small voltage difference, called the input offset voltage, which you can estimate as the output-voltage difference divided by the voltage gain.

▷ How big an input offset voltage do you obtain this way?

### 6.1.5 Common-mode gain

It is desirable for a differential amplifier to be insensitive to *common-mode* input, i.e. identical signals applied to both inputs. This feature (called *common-mode rejection*) is useful when sensing a small signal in the presence of noise, since often the noise is in common on both inputs and can be subtracted away by a differential amplifier.

Test the common-mode rejection of your differential amplifier:

▷ Connect both inputs to the same sine wave from the function generator (i.e. $V_{in+} = V_{in-}$). What do you observe at the outputs? You should see almost identical signals on both outputs. What common-mode gain ($A_{CM}+ = \Delta V_{out}+/\Delta V_{in}+$ and $A_{CM}- = \Delta V_{out}-/\Delta V_{in}-$) do you observe for each output, and why?

Common-mode rejection is usually specified in terms of the *common-mode rejection ratio* expressed in *decibels*:

$$\text{CMRR} = 20 \log_{10} \left( \frac{A_{\text{diff}}}{A_{\text{CM}}} \right). \tag{6.3}$$

▷ What value of CMRR do you observe?

Although the common-mode gain of this circuit is small, it can still be a nuisance in practice. It could be reduced by increasing the size of $R_1$, but that would reduce the amount of current flowing through $Q_1$ and $Q_2$, increasing $r_e$ and reducing the differential gain. A better solution is to replace $R_1$ with a current source (shown in Fig. 6.2).[1] Try this.

▷ Measure how much current your current sink sinks, calculate what you expect, and compare.

---

[1] Strictly speaking, it is a current 'sink' since positive current flows *into* the collector as indicated.

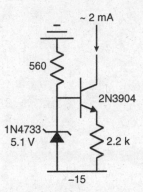

**Fig. 6.2.** Current sink for differential amplifier.

▷ What output signals do you observe now for a common-mode input? What is the common-mode gain now?

It should be *very* much smaller than previously – if the output signal is so small that you have trouble measuring it, you can at least set an upper limit on it and on the common-mode gain.[2]

Save your three-transistor differential amplifier for use below.

## 6.2 Op amps and their building blocks

An operational amplifier is a differential amplifier with a single-ended output and as high a differential gain as possible (typically $>10^5$). Op amps are manufactured as integrated circuits. They are typically used with DC coupling and with *negative feedback* from output to input. Their internal design includes level-shifting circuitry so that the single output is at $\approx 0$ V if the two input voltages are equal.

### 6.2.1 Current mirror

To achieve high gain, in op amps the emitter resistors are typically omitted, and the collector resistors are replaced by current sources. A *current mirror* is a convenient configuration for this purpose.

Build the PNP current mirror of Fig. 6.3. The current out of $Q_4$ 'programs' an approximately equal current out of $Q_5$ as follows: since $Q_4$'s collector is connected to its base, it is held at a $V_{BE}$ drop below the positive supply. This determines the current out of $Q_4$ by Ohm's law applied to $R$.

---

[2] To set an upper limit, assume that the output signal equals the precision of your measurement.

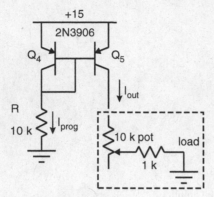

**Fig. 6.3.** Current mirror.

Since the bases are connected together, both transistors have the same value of $V_{BE}$, and thus their collector currents will match if they have matching values of $I_s$ and $\beta$ and are at the same temperature.

In practice, there will always be a slight current mismatch since the programming current includes the base currents of $Q_4$ and $Q_5$ and the output current does not. Also, since $\beta$ increases with $V_{CE}$, and $V_{BE}$ at a given collector current depends slightly on $V_{CE}$ (called the 'Early effect'), the current mismatch will depend on the output voltage.

You can explore this using a variable load, as shown in Fig. 6.3. Monitor the collector voltage as you adjust the load resistance.

▷ How does the output current vary with collector voltage? What is the approximate dynamic resistance of the output? Is this a better or a worse current source than the ones you built in previous chapters?

### 6.2.2 Differential amplifier with current-source loads

Various tricks can be used to improve the performance of the current mirror, as we will see below. But first, hook up your simple current mirror to your three-transistor differential amplifier, replacing the 10 k collector resistors, and remove the 100 Ω emitter resistors (see Fig. 6.4). Connect a scope probe to the collector of $Q_2$. If the inputs balance, the currents through $Q_1$ and $Q_2$ will be equal, and so will the currents through $Q_4$ and $Q_5$. But if a differential signal is present at the input, the current mismatch between $Q_2$ and $Q_5$ must flow through the 10 M scope-probe input impedance. You should then see an enormous differential gain! If $Q_5$ were an ideal current source, the gain would be 10 M/$2r_e$.

▷ Explain the last statement.

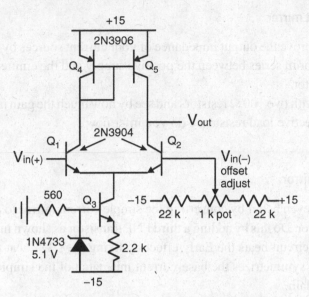

**Fig. 6.4.** Differential amplifier with current-mirror load.

In practice, the collector of $Q_5$ will have an output impedance less than 10 M, and so the gain will be lower.

You may find it desirable to increase the attenuation of your input voltage divider from 100 to 1000 here. Also, to avoid clipping of the output signal, you will want to arrange for the quiescent collector voltage of $Q_2$ to be around 7 V – you can adjust it by hooking up the base of $Q_2$ not to ground, but to the wiper of the 1 k pot, with one end of the pot connected through a resistor to +15 V and the other end connected through a second resistor to −15 V, as shown in Fig. 6.4. Due to the high gain of this amplifier, the output voltage will be sensitive to small variations at either input. If you observe a DC output signal near either ground or +15 V, it is likely that you need to fine-tune the offset voltage using the 1 k pot. Also note that a gain value approaching 1000 is not unreasonable.

▷ What gain do you observe? Compare with the gain you would expect if $Q_2$'s effective load were 10 M. What effective load resistance do you infer? Is this consistent with the dynamic resistance of the current-mirror output you observed in section 6.2.1?

▷ The sensitivity of this circuit is easily demonstrated. Try warming either $Q_4$ or $Q_5$ by gently squeezing the transistor between your finger and thumb. Observe how the output changes and suggest an explanation for your observation. Try doing the same for the other transistor.

### 6.2.3 Improved current mirror

You can improve the output impedance of your current sources by adding a small resistor in series between the positive supply and the emitter of each PNP transistor.

▷ Try this with two 100 Ω resistors and see by how much the gain increases. What effective load resistance do you infer now?

### 6.2.4 Wilson current mirror

You can do even better by converting the simple current mirror to a Wilson current mirror. Do this by adding a third PNP transistor as shown in Fig. 6.5. This clever circuit beats the Early effect by fixing $Q_5$'s $V_{CE}$ – at the same time, it also symmetrizes the base-current mismatch of the simple current mirror. Explain.

▷ Try this to see by how much the gain increases. Note that you will need to re-tune the collector voltage of $Q_2$ by adjusting the offset voltage. What effective load resistance do you infer now?

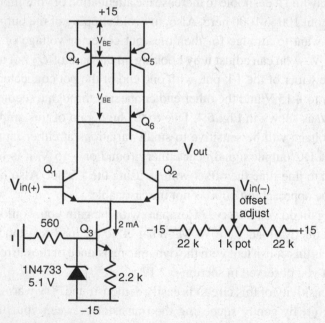

**Fig. 6.5.** Differential amplifier with Wilson-current-mirror load.

IC op amps have even higher gain than this, of course, as well as higher input impedance. Higher input impedance can be achieved by using Darlington transistor pairs in place of the input transistors, or by using FET inputs instead of bipolar transistors. The gain can be increased further by adding a second stage of amplification after the differential pair. To achieve low output impedance, the output is usually buffered with an additional transistor stage.

# 7 Introduction to operational amplifiers

An operational amplifier is a high-gain DC-coupled amplifier with differential inputs and single-ended output. Op amps were originally developed as vacuum-tube circuits to be used for analog computation. Nowadays they are packaged as integrated circuits (ICs). Such devices can closely approximate the behavior of an ideal amplifier, and their use avoids the necessity of coping with the messy internal details of amplifier circuitry. Thus, an IC op amp is often the device of choice in scientific instrumentation. In this chapter we will introduce the op amp and its most common applications.

## Apparatus required

Breadboard, oscilloscope, multimeter, two 741 op amps, one further 741 (optional), one 100 $\Omega$, three 10 k, two 100 k, one 1 M $\frac{1}{4}$ W resistors, and four more 10 k resistors (optional).

## 7.1 The 741 operational amplifier

The IC we shall be using is a general purpose op amp designated by the number 741. The 741 is a very popular and successful design, useful for signals from DC to beyond audio frequency (though in recent years FET-input op amps such as the LF411 have been gaining on the 741 in popularity). It is available from most manufacturers of *linear* integrated circuits (chips that produce an output proportional to their inputs, as opposed to *digital* ICs, whose outputs have typically only two states).

Each manufacturer has a different system of nomenclature for ICs, e.g. National Semiconductor calls the 741 an LM741, Fairchild a μA741, etc., but the 741s made by different manufacturers are all electrically compatible. To add complication, the 741 is available in various package styles and is rated for use in various temperature ranges. The one we use is the

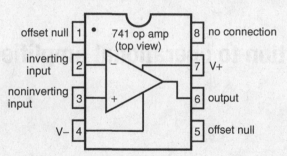

**Fig. 7.1.** Diagram of eight-pin DIP 741 package showing 'pinout'. Often, in addition to (or instead of) the notch at the 'pin 1' end of the package, there is a dot next to pin 1.

741C (commercial temperature range, 0 to 70°C) in the eight-pin mini-DIP (dual in-line pin) plastic package. This package is convenient because the two rows of pins easily straddle the groove running down the center line of a breadboard socket block.

    You can find the manufacturer's data sheet for the 741 on the web or in just about any electronic textbook or linear-IC data book. As shown on the data sheet, the 741 actually has on its single silicon crystal a multiple-stage two-input DC amplifier consisting of twenty transistors, eleven resistors, and one capacitor.[1] Its output voltage is proportional to the voltage difference between the two inputs, i.e. it is a 'differential' amplifier. A differential amplifier is convenient when one wants to study the difference between two almost identical signals, as well as in many other applications, as we shall see. (Of course, this is not to be confused with a differentiator! Recall that a differentiator outputs the time derivative of its input signal, not at all what a differential amplifier does.)

### 7.1.1 741 pinout and power connections

Fig. 7.1 shows the layout of the eight-pin DIP package. The package is a rectangular piece of black plastic containing the silicon chip itself as well as the fine gold wires that connect the chip to the contact pins. Look at the top face of the package and orient yourself as to which pin is pin 1 – looking at the top, you see the pins bending away from you and the pin numbers increasing in the counterclockwise direction. The end at which pin 1 is located is indicated by a special mark – depending on the manufacturer, not necessarily the mark shown in Fig. 7.1. Pins 1–4 are located along one long edge of the package, while pins 5–8 are on the opposite edge.

---

[1] Nowadays, of course, this is nothing – for example, the Pentium chip had 3.3 million transistors.

The 741C is rated for maximum supply voltages of ±18 V, and the recommended range is ±(≤15) V. To be on the safe side, before you begin to build your circuit, turn on the breadboard power and adjust the power supplies to +15 V and −15 V.

Next, turn off the power and insert the op amp into the breadboard, straddling the central groove of a socket block, with pins 1–4 toward the left and 5–8 toward the right. Note that the pins are delicate and are easily bent or broken. If they are too bent to plug into the sockets, straighten them carefully, preferably using needle-nose pliers. Run wire jumpers from the V+ pin to the +15 V bus and from the V− pin to the −15 V bus. Any point of your circuit that is to be 'grounded' should be attached to the common (or ground) bus.

---

**Note:** In the circuits shown below, the pin numbers have been omitted. It is good practice for you to write in the pin numbers yourself before hooking up the circuits, to reduce the possibility of confusion. Trying to work out pin numbers 'on the fly' and keeping them in your head instead of writing them down is a common cause of errors in hooking up circuits.

   **Also note:** Even though the power connections are usually not shown on schematic diagrams of op amp circuits, the positive and negative supplies must always be connected, or the op amp won't do anything!

---

### 7.1.2 An ideal op amp

For a (hypothetical) ideal op amp, zero volts between the *inverting* (marked '−' in schematics) and *noninverting* (marked '+') inputs would yield zero volts at the output (relative to ground); in other words

• the *common-mode gain* and *DC offset* of an ideal op amp are zero.

Moreover,

• the differential gain and input impedance of an ideal op amp are infinite and the output impedance is zero;

• the *bandwidth* (frequency range over which the op amp can correctly operate) and *slew rate* (rate at which the output voltage can change) of an ideal op amp are infinite.

While, of course, it is impossible to build a circuit having these ideal characteristics, one can come remarkably close. The above statements are approximately true for practical op amps – for example, one can buy op amps with gains of $10^5$ to $10^6$.

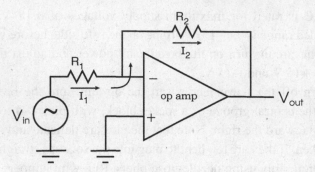

**Fig. 7.2.** Op amp inverting-amplifier circuit. Note the negative feedback resulting from the resistor that connects the output to the inverting input. Op amps are almost always used with negative feedback.

We shall see next that these approximations lead to a very simple way of analyzing op amp circuits.

### 7.1.3 Gain of inverting and noninverting amplifiers

Fig. 7.2 shows an op amp configured as an inverting amplifier. The key principle at work in this circuit is *negative feedback*. The idea is that a fraction of the output signal is applied at the inverting input. Since the gain of the op amp is large and the noninverting input is grounded, any nonzero voltage at the inverting input will cause a large output voltage of the opposite sign. If you think about it, you will see that the only stable situation that can result is that the voltage difference between the inverting and noninverting inputs is zero. In other words, the op amp will do whatever is necessary to zero the voltage difference at its inputs.

Once this principle is grasped, it is easy to compute the gain of the op amp inverting amplifier. Assuming the input currents of the op amp are zero, all the current flowing in through $R_1$ must flow out through $R_2$, i.e. $I_2 = I_1$. Assuming that the *open-loop* voltage gain (i.e. that without any feedback) of the op amp is infinite, the voltage difference at the op amp's inputs must be zero. Applying Ohm's law to $R_1$ and $R_2$, and designating the voltage at the inverting input as $V_-$,

$$V_- = V_{in} - I_1 R_1 = 0 \qquad (7.1)$$

$$\Rightarrow I_1 = -\frac{V_{in}}{R_1} \qquad (7.2)$$

$$V_{\text{out}} = V_- - I_2 R_2 = -I_1 R_2 \tag{7.3}$$

$$\Rightarrow V_{\text{out}} = -\frac{R_2}{R_1} V_{\text{in}}. \tag{7.4}$$

Thus, the *closed-loop* voltage gain (i.e. the gain *with* feedback) of this circuit is

$$A_{\text{v}} = \frac{V_{\text{out}}}{V_{\text{in}}} = -\frac{R_2}{R_1}. \tag{7.5}$$

In a nutshell, since all of the current due to the input signal flows *around* the op amp, the output voltage is determined entirely by Ohm's law applied to $R_1$ and $R_2$. Thus, if $R_1 = 10\,\text{k}$, a gain of $-10$ can be achieved by choosing $R_2 = 100\,\text{k}$, and a gain of $-1$ results from choosing $R_2 = 10\,\text{k}$.

Fig. 7.3 shows an op amp configured as a noninverting amplifier. Again, neglecting the tiny input currents of the op amp and assuming infinite open-loop gain, application of Ohm's law shows that the closed-loop voltage gain of this circuit is

$$A_{\text{v}} = \frac{I_1 R_1 + I_2 R_2}{I_1 R_1} = 1 + \frac{R_2}{R_1}. \tag{7.6}$$

As above, Eq. 7.6 follows from the assumptions that all of the current flows around the op amp and that feedback forces the inverting input to follow the signal applied to the noninverting input. So if $R_1 = 10\,\text{k}$, a gain of 11 can be achieved by choosing $R_2 = 100\,\text{k}$, while a gain of 2 results from choosing $R_2 = 10\,\text{k}$.

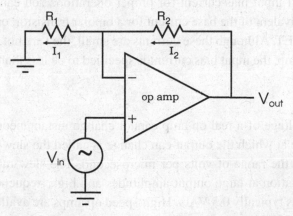

**Fig. 7.3.** Op amp noninverting-amplifier circuit.

### 7.1.4  Op amp 'golden rules'

We see from the above that to understand to a reasonable approximation an op amp circuit with negative feedback, you need only two simple rules.
1. The inputs draw no current.
2. The output does whatever is necessary to maintain the two inputs at equal voltages.

### 7.1.5  The nonideal op amp

Although the ideal op amp approximations are very close to reality, in most situations, here are a few limitations that you should consider:

#### Input offset voltage

A small DC output voltage usually results even when the inputs are identical. The input offset voltage refers to the DC voltage that must be applied at the input to achieve exactly zero volts at the output. The nonzero offset arises due to manufacturing limitations; however, most op amps have pins that allow for external adjustment of the offset. The input offset voltage must be considered when designing small-signal or high-gain circuits. For the 741C op amp, the input offset voltage is specified to be less than 6 mV.

#### Input bias current

An ideal op amp would draw no current at its inputs; however, real op amps require a small input bias current for proper operation. You can think of this as the equivalent of the base current for a bipolar transistor, or the gate current of a JFET. Although these currents are small, they are not zero. For the 741C op amp, the input bias current is specified to be less than 500 nA.

#### Slew rate

The output voltage of a real op amp cannot change instantaneously. The maximum rate at which the output can change is called the slew rate, and is typically in the range of volts per microsecond. The slew rate can be a serious limitation at large output amplitudes and high frequencies. The 741 slew rate is typically 0.5 V/μs. High-speed op amps are available with slew rates of 2000 V/μs.

## 7.2 Experiments

### 7.2.1 Testing open-loop gain

First, try a 741 in the open-loop circuit shown in Fig. 7.4. (As mentioned above, open-loop means that there is no feedback connection between the output and input; op amps are never actually used in this way.) The 1000-to-1 attenuator in the input circuit means that, by adjusting the pot, you can vary the input between $-15$ and $+15$ mV. Try to adjust the input so that the output sits near zero volts.

▷ Vary the input in steps of 100 or 200 $\mu$V over a range that causes the output to vary from its maximum negative voltage to its maximum positive voltage, taking several readings of input and output voltage as you do so. (You may find that the output switches states within a single step!) Approximately what input voltage would result in an output voltage near zero? This is the input offset voltage of your 741.

▷ What are the op amp's positive and negative output saturation voltages (i.e., the maximum voltages it can output)?

If your chip has particularly high gain, you may find that *no* setting of the input voltage causes the output to sit near zero volts; however, you can still set upper and lower limits on the input offset voltage by determining at what input voltages the output switches between its negative and positive saturation voltages. Since your applied input voltages are so tiny, the output voltage is very susceptible to input noise, and you may also observe substantial drift of the output voltage with time. Still, you should

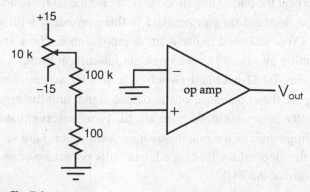

**Fig. 7.4.** Open-loop op amp test circuit.

be able to estimate the gain roughly, or at least place a lower limit on it.

▷ What gain do you observe? Is it consistent with the manufacturer's specification 'gain (typical) = 200 000'?

You may find that stability and noise are improved by keeping the connections between the noninverting input, the 100 Ω resistor, and the inverting input as short as possible – long wires in this current path act as an antenna and pick up electromagnetic noise from the environment, as you will soon observe.

### 7.2.2 Inverting amplifier

Now hook up the inverting amplifier of Fig. 7.2, with $R_1 = 10$ k and $R_2 = 100$ k. Using a 1 kHz sinusoidal input signal with an amplitude of approximately 1 V, look at the input and output.

▷ Verify that the amplifier inverts. Measure the gain and compare with what you expect. Replace the 100 k feedback resistor with 10 k – what gain do you predict and observe now? (Put the 10 k resistor back for the following parts.)

Due to feedback, the output impedance of the inverting amplifier is extremely small: the ≈75 Ω open-loop output impedance of the 741 is multiplied by the ratio of the closed-loop gain to the open-loop gain, here a factor of order $10^{-5}$.

▷ Confirm the small size of the closed-loop output impedance by adding a 100 Ω load to ground at the output. To see that this gives sensitivity to the output impedance, analyze the voltage-divider circuit consisting of the output impedance in series with the load resistance.

▷ Do you expect the output amplitude to decrease measurably under load? How big a decrease do you predict? Is this consistent with what you observe? (You will need to use a small input signal – how small? – to avoid running into the 741's ≈25 mA output-current limit.)

Remove the 100 Ω load and switch to a 10 kHz square wave as the input. Compare the appearance of the output signal and the input signal. For sufficiently large amplitude, you are likely to observe that the leading and trailing edges of the output square wave are not quite vertical.

▷ Measure the slope of the leading edge in volts per microsecond. This is the *slew rate* of the 741C.

According to the manufacturer, the typical slew rate is 0.5 V/μs. There exist more expensive op amps with much higher slew rates.

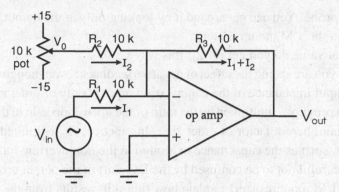

**Fig. 7.5.** Circuit for demonstrating a summing junction. Since the inverting input is held near ground due to feedback, $V_{\text{out}} = (I_1 + I_2) \cdot R_3$, where $I_1 = V_{\text{in}}/R_1$ and $I_2 = V_0/R_2$. If $R_1 = R_2 = R_3$, $V_{\text{out}} = V_{\text{in}} + V_0$.

The inverting input is called a 'virtual' ground because it is kept at zero volts by feedback. Thus, even though the open-loop input impedance of the inverting input is very large, it acts here like a short circuit to the noninverting input (ground). Since the input signal sees $R_1$ to ground, the input impedance of the inverting amplifier is equal to the value of $R_1$.

Note that the virtual ground at the inverting input can also be used as a 'summing junction': all currents arriving at that point are summed and passed through the feedback resistor to the output. To demonstrate this, feed an adjustable current into the summing junction, by adding a 10 k pot with 10 k resistor in series with the wiper, as shown in Fig. 7.5, and examine the amplifier output for a small-amplitude 1 kHz sinusoidal input.
▷ What happens to the output DC offset as you adjust the pot, and why?

### 7.2.3 Noninverting amplifier

Set up the noninverting amplifier circuit of Fig. 7.3 with $R_1 = 10$ k.
▷ With a 1 kHz sinusoidal input, measure the gain with $R_2 = 100$ k and with $R_2 = 10$ k. Compare with what you expect. Verify that the amplifier is noninverting. (Leave the 100 k resistor in for the following parts.)
▷ Try to measure the input impedance by putting a 1 M resistor in series with the input and looking at the signal before and after the resistor. Explain your result.

The input impedance can be inferred by analyzing the voltage-divider circuit consisting of the 1 M resistor in series with the input impedance. If your answer is 10 M, consider that this is the input impedance of the attenuating scope probe! You can get around it by looking only at the output, with and without the 1 M input resistor.

▷ What value do you get for $Z_{in}$ this way?

Again you are seeing the effect of negative feedback: even though the open-loop input impedance of the noninverting input is only of order megohms, it is, in principle, multiplied by the ratio of the open-loop gain to the closed-loop gain, here a factor of order $10^5$. (In practice, $Z_{in}$ is limited by other effects, such as the capacitance to ground at the noninverting input.)

▷ Be careful not to be confused by the DC shift in the output produced by the 1 M input resistor! Explain how this shift results from the op amp's input bias current. What value for the input bias current is implied by the observed DC shift?

The output impedance should of course be the same as for the inverting amplifier, since, as far as output impedance is concerned, it is the same circuit!

▷ Explain this last statement. (Hint: what was the only thing you had to change to make the noninverting amp from the inverting amp?)

### 7.2.4 Voltage follower

As shown in Fig. 7.6(a), connect the output of the op amp directly to the '−' input, and connect the output of the function generator to the '+' input of the op amp.

▷ After turning on the power, confirm that the input and output signals are identical and that the voltage follower is noninverting. Record the input and output amplitudes.

▷ Of what possible use is an amplifier of unity gain that does not even invert the signal?

As an illustration of the voltage follower's usefulness, coil a long wire (30 to 50 cm in length) around the outside of an AC power cord. (Any power cord will do provided that it is plugged in!) The breadboard cord is usually the most convenient. Leave one end of your coil floating, and connect the other to any convenient spot on the breadboard.

▷ Using your scope probe, observe the waveform. Record and explain what you see.

▷ Given that the scope probe has an input impedance of 10 MΩ, estimate the power of this signal (i.e., how much power is dissipated by the scope probe?).

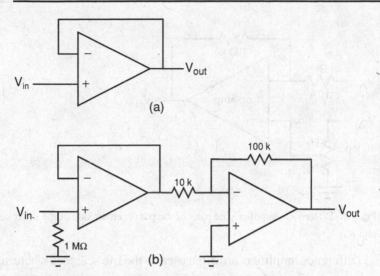

**Fig. 7.6.** (a) Op amp voltage follower; (b) voltage follower as the input stage to an inverting-op-amp circuit.

You can reproduce this effect more simply by touching the probe tip with your finger. 60 Hz noise is pervasive throughout North America (50 Hz in Europe) and is often the dominant background noise in electronic equipment.

▷ Try amplifying this low-power signal using the inverting-amplifier circuit previously constructed. Sketch the output and record your observations.

▷ Now, instead of driving the amplifier directly, insert a voltage follower as shown in Fig. 7.6(b). Record the follower output as well as the amplifier output. If the amplifier output saturates, choose a smaller feedback resistor to reduce the gain of the inverting amp. Explain your observations.

## 7.2.5 Difference amplifier

Fig. 7.7 shows the 741 configured as a difference amplifier, with the output voltage equal to the difference of the two input voltages. A difference amplifier is both an inverting amp and a noninverting amp. An inverting amp is created if $V_{in+}$ is grounded, whereas a noninverting amp is created if $V_{in-}$ is grounded. If both inputs are connected to signals, the difference between the two signals is amplified. This follows since, if these inverting and noninverting amplifiers have equal gains, the output is proportional to the difference of the inputs. The gains are matched provided that $R_1/R_2 = R_3/R_4$.

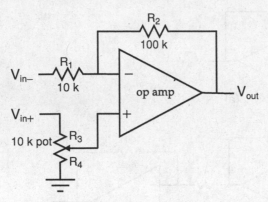

**Fig. 7.7.** Difference amplifier. The parts of the pot on either side of the slider serve as $R_3$ and $R_4$.

Difference amplifiers are often used in the life sciences where signals are small and exist within a noisy environment, e.g. in the electrocardiograph (ECG). Any background noise common to both inputs (*common-mode noise*) is rejected, while the signal of interest (present at only one of the inputs) is amplified and appears at the output. A potentiometer is often used (as in Fig. 7.7) to tune the gain and common-mode rejection of the amplifier.

The quality of the amplifier is measured (in part) by the common-mode rejection ratio (CMRR), based on the ratio of the differential voltage gain and common-mode voltage gain:

$$CMRR = 20 \log \left( A_{diff} / A_{CM} \right). \tag{7.7}$$

It is customary to give the CMRR in decibels, as shown in Eq. 7.7. The 741 general-purpose op amp is a differential amplifier with a CMRR value specified to be at least 70 dB. Precision op amps are commercially available with CMRR values as high as 140 dB or more.

Build the circuit shown in Fig. 7.7. As always, be careful not to short the potentiometer's center tap to ground or power. You can estimate the common-mode rejection ratio by measuring both the common-mode voltage gain and the differential voltage gain.

The common-mode voltage gain is determined by applying identical signals to both inputs and observing the output voltage: $A_{CM} = V_{outCM} / V_{inCM}$. Using a 1 kHz sine wave at maximum amplitude as your input, tune your difference amp to minimize the output amplitude (i.e. adjust the potentiometer until the output amplitude is as small as possible). The best estimate for the common-mode gain can be made using the averaging feature from the

'ACQUIRE' menu of the TDS210 oscilloscope. It may be possible, depending on the quality of your op amp, to tune the potentiometer such that no output signal is visible, even at the most sensitive scale.

▷ Verify that $R_1/R_2 = R_3/R_4$.

▷ Estimate (or set an upper limit on) the common-mode voltage gain.

The resistance ratios can be measured using your meter; however, be sure to turn off the power and disconnect each component before making your measurements – accurate resistance measurements **cannot** be guaranteed if the component remains connected to the circuit, since then you are measuring a parallel combination with the other components rather than just the resistance of interest. For example, when measuring the resistances between the potentiometer center tap and ends, be sure to disconnect all wires connecting the pot to the input signal, op amp, and ground. Replace all resistors and wires when done.

To measure the differential voltage gain, ground one input while applying a 1 kHz sine wave of amplitude 1 V or less at the other input. It shouldn't matter which input you ground. If you're curious, try it both ways and see if you get equal gains.

▷ Measure the differential voltage gain.

▷ Explain why it doesn't matter which input is grounded when measuring the differential voltage gain.

▷ Estimate the CMRR for this circuit.

▷ What is the input impedance at each input?

## 7.3 Additional experiments

### 7.3.1 Current source

An ideal current source would maintain a constant current through the load, regardless of load resistance. Try the op amp current source of Fig. 7.8. Vary the 'load' pot as you measure the load current with a digital multimeter and the load voltage with a multimeter or scope; what do you observe?

▷ Explain how this circuit works (use diagrams and equations as necessary).

▷ What should the current be, and why?

▷ What is $dI/dV$? What is the 'compliance' (output-voltage range over which $dI/dV$ is small) of your current source?

▷ Compare the performance of this op amp current source to the transistor current sources you built previously.

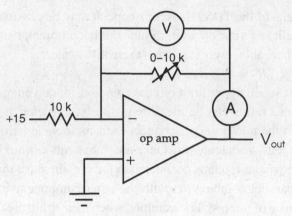

**Fig. 7.8.** Op amp current source.

## 7.3.2 Noninverting summing amp with difference amplifier

This circuit is fun to build and to observe in action. It also serves as an excellent demonstration of the constructive and destructive interference between two waveforms.

As stated above, difference amplification is usually used to eliminate common-mode noise. The circuit shown in Fig. 7.9 sums two waveforms and then eliminates one using the difference amp. One waveform will be from the function generator while the other will be a 60 Hz wave created as in section 7.2.4. (If two function generators are available, feel free to replace the voltage follower with the output from the second generator.)

First construct the voltage follower and summing amplifiers as shown in Fig. 7.9. Set the function generator to a sine wave with a frequency between 59 and 61 Hz.[2] Adjust the generator amplitude to match the amplitude of the voltage follower and observe the output of the summing amp. The output should be the linear sum of the two input waveforms.

Have fun with the circuit by changing the amplitude and frequency. Observe what happens with either a triangle- or square-wave input. Using the 'AC line' as your trigger source may be useful here.

Now add the difference amplifier. With the function-generator amplitude set to zero, adjust the potentiometer until the difference amp output has zero amplitude (as with the difference amplifier previously built). Now increase the amplitude of your function generator and observe the difference

[2] If the local supply frequency is not 60 Hz, set the function generator frequency to be within 1 Hz of your local value.

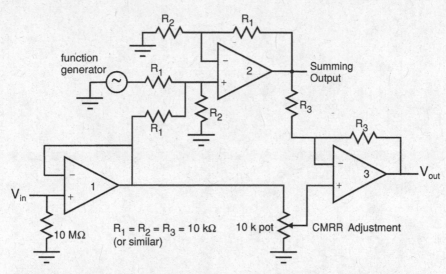

**Fig. 7.9.** Fancy summing circuit. Amp 1 is a voltage follower used to buffer the 60 Hz pickup on a wire wrapped around an AC power cord. Amp 2 is a noninverting summing amplifier with unity gain. Amp 3 is a difference amplifier with an adjustment to maximize the CMRR.

amplifier output. Be sure to switch your trigger source back to the appropriate input channel. Try changing the input frequency. Replace the 60 Hz AC-line signal at the noninverting input of the difference amp with the output from the function generator. Observe how the output changes.

▷ Explain how this circuit works using diagrams and equations as needed. Explain why the summing amp isn't inverting and why it has unity gain.

▷ Sketch the inputs and output of the summing amp for a function generator frequency near 60 Hz. Why doesn't the output have a well defined amplitude?

▷ Sketch the inputs and output of the difference amplifier. How does the output change when the inputs are switched?

▷ Is the output inverted with respect to the original inputs? If so, why? If so, what could you change to make the output of the difference amp noninverted with respect to the original inputs?

# 8 More op amp applications

In Chapter 7 we studied some of the basic properties of operational amplifiers. There are an enormous number of ways that op amps can be applied to process analog signals. In this chapter we will explore several such applications: circuits that differentiate or integrate their input voltage as a function of time, form the logarithm or exponential of their input voltage, or rectify their input voltage. The op amp versions of these applications come closer to the ideal than the passive versions of some of them that you studied in earlier chapters. We will also see how to use feedback to compensate for the limitations of discrete devices.

## Apparatus required

Breadboard, dual-trace oscilloscope with two attenuating probes, one 741C and one LF411 operational amplifier, one 1 k, two 10 k, and one 100 k $\frac{1}{4}$ W resistor, 0.0047 μF and 0.033 μF capacitors, two 1N914 (or similar) silicon signal diodes, 2N3904 and 2N3906 transistors.

## 8.1 Op amp signal processing

Recall that for an inverting amplifier made from an op amp, with input resistor $R_i$ and feedback resistor $R_f$, the gain is $-R_f/R_i$ (neglecting the input offset voltage and offset and bias currents and taking the op amp open-loop gain to be infinite). We can generalize this result for devices other than resistors, as illustrated in Fig. 8.1.

$$A_v = -\frac{Z_f}{Z_i}. \tag{8.1}$$

Eq. 8.1 is useful if we are analyzing circuit performance in the frequency-domain for a sine-wave input, but often we are concerned with the response

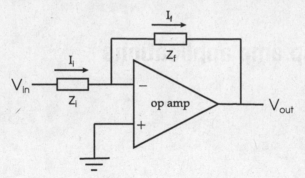

**Fig. 8.1.** Generalized op amp inverting-amplifier circuit.

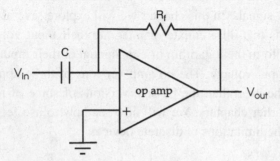

**Fig. 8.2.** Basic op amp differentiator.

in the time-domain to an arbitrary input waveform. Then we can analyze the circuit using Ohm's law. Since, to an excellent approximation, $I_f = I_i$, the gain is determined by the current–voltage characteristics of the input and feedback devices, as we will see in more detail below.

### 8.1.1 Differentiator

As shown in the circuit of Fig. 8.2, the basic op amp differentiator (not to be confused with the difference amplifier) is similar to the basic inverting amplifier studied in Chapter 7, except that the input element is a capacitor rather than a resistor. Using the assumption that the output does whatever necessary to maintain the two inputs at equal voltages, it is easy to show that the output voltage is given by

$$V_{out} = -I R_f = -\frac{dQ}{dt} R_f = -R_f C \frac{dV_{in}}{dt}, \tag{8.2}$$

since $Q = C V_{in}$, where $Q$ is the charge stored on the capacitor.

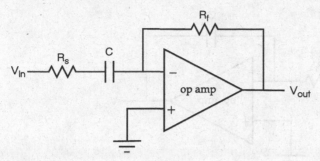

**Fig. 8.3.** Improved op amp differentiator.

One problem with the basic circuit is that the capacitor's reactance,

$$X_C = \frac{1}{2\pi f C},\tag{8.3}$$

decreases with increasing frequency. Since here $Z_i = X_C$, Eq. 8.1 shows that the output voltage of the basic differentiator increases with frequency, making the circuit susceptible to high-frequency noise and prone to oscillation.

A more practical differentiator circuit is shown in Fig. 8.3, with a resistor placed in series with the input capacitor to limit the high-frequency gain to the ratio $R_f/R_s$. The output voltage as a function of time is still given by Eq. 8.2, as long as the input frequency is small compared with

$$f = \frac{1}{2\pi R_s C}.\tag{8.4}$$

For input frequencies greater than this, the performance of the circuit approaches that of an inverting amplifier with voltage gain

$$A_v = -\frac{R_f}{R_s}.\tag{8.5}$$

### 8.1.2 Integrator

By interchanging the resistor and capacitor in the differentiator circuit of Fig. 8.2, we obtain an op amp integrator. As shown in Fig. 8.4, the resistor $R_i$ is the input element and the capacitor $C$ is the feedback element. The output voltage, as a function of time, is given by

$$V_{out} = -\frac{1}{R_i C} \int V_{in} dt,\tag{8.6}$$

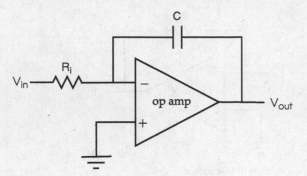

**Fig. 8.4.** Basic op amp integrator.

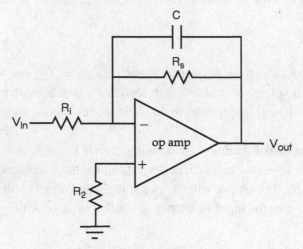

**Fig. 8.5.** Improved op amp integrator.

which is proportional to the time integral (area under the curve) of the input waveform vs. time.

As in the case of the differentiator, a more practical integrator circuit is shown in Fig. 8.5. The resistor $R_s$ across the feedback capacitor, called a 'shunt resistor', is used to limit the low-frequency gain of the circuit. If the low-frequency gain were not limited, the input DC offset, although small, would be integrated over the integration period, possibly saturating the op amp.

To minimize the DC offset voltage resulting from the input bias current, the resistor $R_2$ should equal the parallel combination of the input and shunt resistors:

$$R_2 = \frac{R_i R_s}{R_i + R_s}. \tag{8.7}$$

Since the shunt resistor limits the circuit's low-frequency gain, Eq. 8.6 is valid for input frequencies greater than

$$f = \frac{1}{2\pi R_{s} C}. \tag{8.8}$$

For input frequencies less than $f$, the performance of the circuit approaches that of an inverting amplifier with voltage gain

$$A_{v} = -\frac{R_{s}}{R_{i}}. \tag{8.9}$$

### 8.1.3 Logarithmic and exponential amplifiers

By using a diode as the input or feedback element, we obtain a circuit that takes the logarithm (Fig. 8.6) or exponential (Fig. 8.7) of its input signal. For the log amplifier, we can analyze the circuit performance as follows; the analysis of the exponential amplifier is left as an exercise.

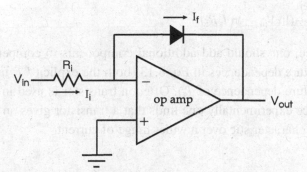

**Fig. 8.6.** Op amp logarithmic amplifier.

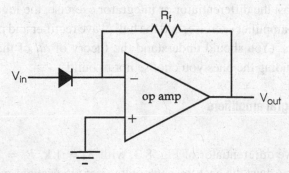

**Fig. 8.7.** Op amp exponential amplifier.

The current $I_f$ through the feedback element equals the input current $I_i$, which is determined by the input voltage $V_{in}$:

$$I_f = I_i = \frac{V_{in}}{R_i}. \tag{8.10}$$

We can relate this to the output voltage $V_{out}$ using the exponential diode current–voltage law:

$$I_f = I_s \left( e^{-eV_{out}/nkT} - 1 \right). \tag{8.11}$$

The minus sign in the exponential reflects the fact that the anode of the diode is connected to virtual ground; thus, for $I_f$ positive, $V_{out}$ is negative. The constant $n$ has been introduced since (as we saw in Chapter 3) the slope of the exponential for a silicon diode is not quite as steep as $e/kT$; one finds experimentally that $n \approx 2$ for silicon diodes and $n \approx 1$ for germanium. Thus the output voltage is

$$V_{out} \approx -\frac{nkT}{e} \ln \frac{I_i}{I_s} \tag{8.12}$$

$$\approx -\frac{nkT}{e} (\ln V_{in} - \ln I_s R_i). \tag{8.13}$$

In practice, one should add additional components to compensate for the temperature dependences in Eq. 8.13 (both the explicit $kT$ factor and the temperature dependence of $I_s$). Often, a transistor is used in place of a diode, since experimentally one finds that a transistor gives an accurate exponential characteristic over a wider range of current.

## 8.2 Experiments

Complete at least the differentiator or integrator exercise, the logarithmic- or exponential-amplifier exercise, and the half-wave rectifier and push–pull driver exercises. (You should understand the theory of *all* of the circuits discussed, including the ones you choose not to build.)

### 8.2.1 Differential and integral amplifiers

#### Differentiator

Set up the active differentiator of Fig. 8.3, with $R_s = 1$ k, $R_f = 10$ k, and $C = 0.033$ μF. Adjust the peak-to-peak voltage of the triangle-wave input to 10 V, and use a frequency in the neighborhood of 500 Hz.

▷ Measure carefully the peak-to-peak voltage $V_{\text{p-p}}$ and period $T$ of the input signal, and from them compute the slope of the input voltage vs. time,

$$\frac{dV_{\text{in}}}{dt} = \frac{2V_{\text{p-p}}}{T}. \tag{8.14}$$

Now look at the function-generator output on channel 1 of the scope and the op amp output on channel 2, while triggering on channel 1. You should see that the op amp output signal is a square wave that is 90° out of phase with the input, i.e., the output signal is a representation of the negative of the time derivative of the input.

▷ Measure the step height in volts of the square-wave output of the op amp.

The theoretical prediction is

$$\text{step height} = 2R_f C \left| \frac{dV_{\text{in}}}{dt} \right|. \tag{8.15}$$

Now change the input frequency from 500 Hz to 10 kHz. Be sure to reduce the input amplitude to avoid saturating the output voltage. Record the appearance of the output signal at this frequency. Measure the peak-to-peak output voltage and determine the voltage gain.

▷ Sketch the input and output waveforms at 500 Hz and 10 kHz and comment on your results.

▷ Compare your data with Eq. 8.15 and compare the measured gain at 10 kHz with the theoretical expectation.

▷ At what approximate frequency does this circuit cease to act as a differentiator (i.e. approach the operation of an inverting amplifier)?

## Integrator

Set up the circuit shown in Fig. 8.5, with $R_i = R_2 = 10\,\text{k}$, $R_s = 100\,\text{k}$, and $C = 0.0047\,\mu\text{F}$. Adjust the peak-to-peak voltage of the square-wave input to 1 V and the frequency to 10 kHz. You should see an output signal that is a triangle wave 90° out of phase with the input square wave.

▷ Derive Eq. 8.6.

▷ Measure the peak-to-peak voltage of the triangle wave and compare with the value you would expect theoretically.

Now change the input frequency to 100 Hz.

▷ Record, describe, and explain the appearance of the output signal at this frequency.

▷ Measure the peak-to-peak output voltage and determine the voltage gain, comparing with what you would expect theoretically.

▷ Sketch the input and output waveforms at 10 kHz and 100 Hz, and comment on your results. At what approximate frequency will this circuit cease to act as an integrator (i.e. approach the operation of an inverting amplifier)?

▷ Compare the output amplitudes with the theoretical expectation.

### 8.2.2 Logarithmic and exponential amplifiers

Set up the circuit of Fig. 8.6 with $R_i = 10$ k or Fig. 8.7 with $R_f = 10$ k. Figure out how to apply an adjustable DC input voltage.

▷ Verify the logarithmic or exponential gain characteristic for several input voltages that probe the full range of output voltage of which the circuit is capable.

▷ Explain how the circuit works and derive an expression that relates the input voltage to the output voltage.

▷ Make a semilog plot of your data and find the experimental value of $n$ for your diode.

▷ The log of zero is undefined and any number raised to the power zero equals one. Why then do the logarithmic and exponential amplifiers give an output of zero volts when the input is zero volts?

### 8.2.3 Op amp active rectifier

Fig. 8.8(a) shows a simple op amp half-wave rectifier.

▷ Build it with $R = 10$ k and try it out with a low-frequency sine-wave input, in the neighborhood of 100 Hz. Record the input and output waveforms.

This circuit suffers from a drawback: there is effectively no feedback during the half of the input-wave cycle when the diode is reverse-biased! (This is obvious since the voltages at the inverting and noninverting inputs are not equal during that time.) In other words, the reverse-biased diode has such a large impedance that the gain becomes very large and the op amp saturates.

▷ To verify this, look at the waveform at the op amp's output pin.

Because of the limited slew rate of the op amp, saturation limits the circuit's performance at high input frequencies. Try a high frequency (say 10 kHz) and record in detail what you see, both at the op amp output and at the rectifier output. Replace the 741 with a higher speed op amp such as the LF411, and repeat your observations. Also, be sure to use a signal

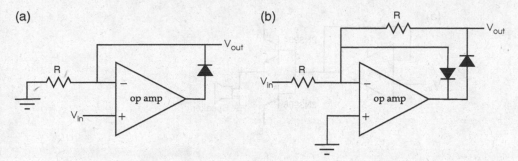

**Fig. 8.8.** (a) Simple op amp half-wave rectifier; (b) improved version.

or switching diode like the 1N914 – while switching diodes are designed for high frequencies, rectifier diodes (e.g., the 1N4001) tend to have large junction capacitance and thus have poor performance at high frequency.

The circuit of Fig. 8.8(b) overcomes the slew rate limitation.

▷ Build it, try it out, and figure out why it has much better response at high frequency. Explain how each circuit works (use diagrams if necessary) and explain why one is better than the other.

▷ Sketch $V_{in}$, $V_{out}$, and the op amp output for both half-wave rectifier circuits at 100 Hz and at 10 kHz.

▷ How is this active rectifier better than a simple diode rectifier? (For example, what would be the response of a simple rectifier to an input signal of amplitude less than 0.7 V? How does this circuit respond to such a signal?)

▷ Comment on the performance of the LF411 as compared with that of the 741.

### 8.2.4 Op amp with push–pull power driver

A typical op amp (such as the 741) by itself does not have enough output-current capability to drive an 8 $\Omega$ load such as a speaker. The addition of a push–pull driver stage to buffer the output is a common solution. When large output power is needed (several watts) as in audio applications, power Darlingtons are often employed as the push–pull transistors.

Recall that in a previous chapter you used a push–pull power driver to drive the breadboard's speaker from the function generator. This circuit displayed crossover distortion since one transistor switched off $2V_{BE}$ before the other switched on. One solution to crossover distortion is to employ an

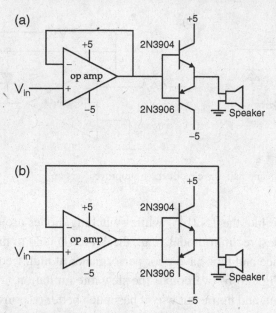

**Fig. 8.9.** Op amp follower with push–pull output-buffer power driver with two feedback arrangements: (a) feedback before and (b) feedback after power driver.

op amp to compare the output signal with the input signal and correct the drive signal to the push–pull stage to compensate for the $2V_{BE}$ gap.

▷ First hook up the circuit of Fig. 8.9(a) to see crossover distortion in action. This circuit is susceptible to noise, so be neat and orderly and keep the leads as short as possible. Use an audio-frequency sine wave, in the vicinity of 1 kHz.

▷ Record the input and output waveforms – how do they differ, and why?

▷ Rearrange the feedback loop to include the push–pull driver inside it (Fig. 8.9(b)), and compare the input and output waveforms again. What does the signal at the op amp output look like, and why?

▷ How much power is dissipated in the speaker assuming a sine wave of amplitude 4 V? How does the peak current through the speaker compare with the 741's maximum output current?

▷ Sketch the op amp input, the op amp output, and the waveform at the speaker for both circuits and explain how the op amp eliminates crossover distortion.

▷ Estimate the total power consumed by this circuit.

▷ What is the maximum input amplitude that can be accurately reproduced at the speaker (without clipping)?

$$V_{out} = V_{in}^{A}$$

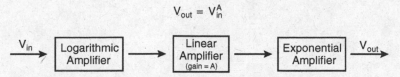

**Fig. 8.10.** Block diagram showing how to build an 'exponentiator': a circuit that creates an output voltage equal to the input voltage raised to any desired power.

## 8.3 Additional exercises

The availability of log and antilog circuits allows elaborate mathematical operations on voltages. For example, the square of a number can be found by taking the logarithm of the number, multiplying by two, and then taking the antilogarithm of the result (see Fig. 8.10); similarly, the product of two numbers equals the antilog of the sum of their logarithms:

$$x^2 = \log^{-1}(2\log x) \tag{8.16}$$

$$x \times y = \log^{-1}(\log x + \log y). \tag{8.17}$$

Choose an arithmetic function of your choice (other than addition, subtraction, and multiplication by a constant) and design a circuit using op amps to perform that function. Make a plot of the output *vs.* input voltages to verify that the circuit works correctly. Discuss the limitations of your circuit.

# 9 Comparators and oscillators

In this chapter you will encounter some applications of positive feedback in op amp and comparator circuits. You will see how uncontrolled feedback can cause unwanted oscillation, and how controlled positive feedback (hysteresis) can be used to eliminate unwanted oscillation or produce intentional oscillation. There is also an optional active-filter application at the end.

## Apparatus required

Breadboard, dual-trace oscilloscope with two attenuating probes, two 741 and one LF411 op amp, 311 comparator, 555 timer, one 100 Ω, one 820 Ω, two 1 k, two 3.3 k, three 10 k, one 100 k, one 1 M, and one 10 M $\frac{1}{4}$ W resistor, three 0.033 μF, one 0.01 μF, and one 1 μF capacitor, one red LED, and two 3.3 V Zener diodes.

## 9.1 Experiments

### 9.1.1 Op amp as comparator

Begin by wiring up a 741 in open-loop mode as you did in a previous lab (Fig. 9.1(a)). With no negative feedback, the saturated output that results allows the op amp to be used as a voltage comparator – a circuit that tells you whether an input voltage is higher or lower than a 'threshold' voltage (the threshold is ground in this case). Since op amps are not specifically engineered for open-loop operation, it is not a very good voltage comparator (in ways that we shall soon see), but in some situations (when high-speed response and high sensitivity are not required) an open-loop-741 'comparator' is perfectly adequate.

▷ Start by applying a 1 kHz sine wave to the input and observe what the circuit is doing.

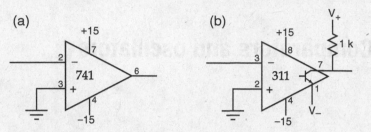

**Fig. 9.1.** (a) Poor comparator: 741 op amp used in open-loop mode; (b) 311 comparator. Pinouts shown for eight-pin mini-DIP package.

▷ Now raise the input frquency to 100 kHz. Notice that the output 'square wave' is not very square. Try an LF411 op amp in place of the 741.

▷ For each case studied, sketch the output waveform and measure the output amplitude and DC offset. Explain your measurements.

▷ Contrast the performance of the LF411 and 741 when configured as comparators. What op amp limitation is responsible for the poor comparator performance at high frequency?

Now substitute a 311 comparator for the op amp. (The pinouts are **NOT** the same; see Fig. 9.1(b).) Whereas op amps are intended to be used with negative feedback, the 311 is specifically designed for operation in open-loop mode or with positive feedback.

The 311's output stage differs from that of an op amp: to enhance flexibility, it has both a positive output (pin 7) and a negative output (pin 1). We shall be using the positive output. In this configuration, a 'pull-up' resistor to a positive supply is required, in order to determine the output-voltage level, and the negative output is generally connected to ground.

Note that the positive output is the collector of an NPN bipolar transistor. As such, it cannot source current, but it *can* sink current. When the output transistor is off (i.e., its base voltage is less than or equal to its emitter voltage), the collector is pulled up to $V_+$ by the external pull-up resistor. Conversely, when the base voltage is raised 0.7 V above the emitter voltage, the transistor saturates and the output pin is pulled down close to the emitter voltage. This output configuration gives the 311 maximum flexibility to provide the various signal voltages used by digital logic chips, including TTL, CMOS, and ECL logic levels.[1]

▷ Start by choosing convenient values for $V_+$ and $V_-$. Use an input frequency of 100 kHz and observe the output.

---

[1] Logic levels are discussed in section 10.1.1.

▷ Change $V_+$ and $V_-$. Record and explain any changes in the output waveform.

▷ How does the output signal from the 311 differ from that of an op amp? What is the 311's slew rate?

## 9.1.2 Unintentional feedback: oscillation

An unfortunate side-effect of the 311's fast response is its tendency to oscillate when presented with a sufficiently small voltage difference between its inputs. The unavoidable capacitive coupling (typically a few picofarads, depending on how the circuit is wired) between the output and the input causes some feedback, so, when the output switches state, a small transient signal is picked up at the input. There is also some feedback due to the changing input base current when one input transistor switches on and the other switches off. These effects can result in self-sustaining oscillation.

The pickup occurs through what is essentially a voltage divider, consisting of the large coupling impedance (i.e., small capacitance) between the output and the input, together with whatever impedance to ground is present at the input. The effect is thus mitigated if there is a low impedance to ground at the input, and exacerbated if the impedance is high.

Try to make your 311 oscillate by feeding it a triangle wave with a gentle slope $dV/dt$ (i.e., small amplitude and/or low frequency). Use $V_+ = +15$ V and $V_- =$ ground. To exacerbate the feedback, insert 10 k in series with the function-generator source resistance as shown in Fig. 9.2. When looking at the input signal, it is a good idea here to use the ground clips on your

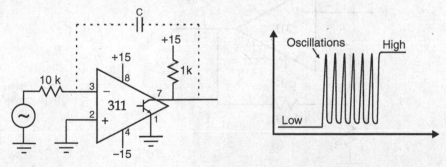

**Fig. 9.2.** 311 comparator with 10 k series input resistor. The capacitor shown connecting the input and output represents the stray capacitance associated with the breadboard. (Do **not** add a discrete capacitor!)

scope probes, to suppress pickup of the large transient pulse due to the 15 V swing of the 311's output. You'll want to look at a low-to-high or high-to-low transition and 'zoom in' by increasing the time sensitivity to look for the rapid multiple transitions that indicate oscillation.

▷ Starting with an input frequency near 100 kHz and an amplitude of several volts, gradually reduce the input amplitude and frequency until oscillations are observed. To be able to see the rapid transitions of the output, be sure the time scale on your oscilloscope is about 500 ns per division or less.

▷ Sketch the observed oscillations. What is the time scale of the oscillations – i.e., what is the period $\Delta t$ for one cycle of the oscillation waveform?

▷ How small an input slope does it take to cause oscillation?

▷ What causes the oscillations to stop?

▷ Why is a 741 configured as a comparator less likely to oscillate than the 311?

### 9.1.3 Intentional positive feedback: Schmitt trigger

The oscillation problem can be eliminated by adding controlled positive feedback (*hysteresis*) as shown in Fig. 9.3. The hysteresis also makes the circuit less sensitive to noise.

The phenomenon of hysteresis should be familiar from your study of magnetic materials in introductory physics. In the present context, the term

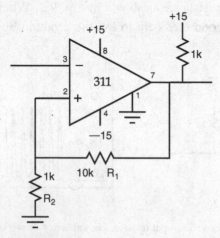

**Fig. 9.3.** Schmitt trigger using 311 comparator.

hysteresis refers both to the situation of having two different comparator thresholds, depending on the history of the input signal, as well as to the size of the voltage difference between the two thresholds. The threshold voltage at the noninverting terminal is found by applying the voltage-divider equation

$$V_{in+} = V_{out}\frac{R_2}{R_1 + R_2}. \tag{9.1}$$

There are two cases to consider, since the output might be at $+15$ V or it might be at ground – i.e., since $V_{out}$ has two possible states, $V_{in+}$ has two possible states.

▷ Build the circuit and apply a signal with an amplitude of at least 1.5 volts. Try to create output oscillations. Notice the rapid and clean transitions at the output, independent of the input waveform or frequency.

▷ Sketch the output waveform. Is the output symmetric? If not, why not?

To see exactly what is happening, use the two-channel oscilloscope to display the voltages at both comparator input terminals simultaneously. Be sure to use DC coupling for both channels, and set them to the same voltage sensitivity and the same zero offset.

▷ Carefully sketch both input waveforms on the same graph and explain how the circuit works. How does the hysteresis prevent oscillations?

▷ Based on the component values used, what do you predict for the comparator thresholds? What is the predicted amount of hysteresis in volts? How do these compare with your observations?

▷ Record what you see as you vary the input amplitude. Below some input amplitude the output stops switching. At what amplitude does this occur, and why?

### 9.1.4 *RC* relaxation oscillator

By adding an *RC* network to your Schmitt trigger, construct a square-wave oscillator, as shown in Fig. 9.4. Since a fraction of the output is fed back into both the inverting and noninverting inputs, this circuit has both negative and positive feedback. By connecting pin 1 to $-15$ V as shown, you should get a symmetrical square-wave output with little DC offset. No input from the function generator is required; the circuit should oscillate spontaneously.

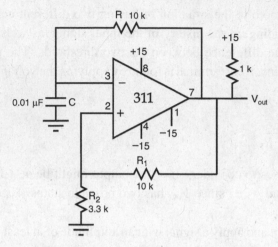

**Fig. 9.4.** $RC$ relaxation oscillator using comparator.

The expected frequency of oscillation is

$$f = \frac{1}{2RC \ln\left(\frac{2R_1}{R_2} + 1\right)}. \tag{9.2}$$

▷ Sketch the output waveform and explain how this circuit works. (Hint: it operates by repeatedly charging and discharging the capacitor between the two threshold voltages of the Schmitt trigger.)
▷ What are the threshold voltages?
▷ Why does the circuit oscillate spontaneously?
▷ Derive Eq. 9.2 and compare with the observed frequency.

### 9.1.5 555 timer IC

Nowadays, one hardly ever builds a square-wave oscillator from an op amp or comparator, since the 555 series of timer chips makes designing stable, predictable oscillators easy. Note that a 555 is not an op amp or comparator, but an oscillator 'kit', including two comparators, among other items (see Fig. 9.5). The 555 can also be used as a timer, as we shall see below.

The 555 is an eight-pin IC powered from a positive voltage source. The supply voltage can range from 4.5 to 16 V. As with a comparator, the output is 'digital', with a high value near $V_{CC}$ and a low value near ground. The output sinks current while low and sources current while high (up to 200 mA in either case). Also, as with a comparator, the 555 output slew rate

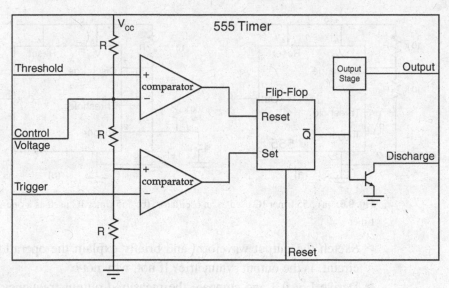

**Fig. 9.5.** Block diagram for the 555 timer IC.

is high, and the 555 can change states within 100 ns. Refer to a 555 data sheet for additional details.

To see how the 555 works, first examine the connections associated with the two comparators. The top comparator output is high when the threshold input is greater than $\frac{2}{3}V_{CC}$, and low otherwise. The bottom comparator output is high when the trigger input is less than $\frac{1}{3}V_{CC}$, and low otherwise. In general, TRIGGER and THRESHOLD should be configured such that only one of the comparators is high at any given moment. The outputs connect to a flip-flop (described in detail in Chapter 11). A positive signal at SET causes $\overline{Q}$ to be near ground, while a positive signal at RESET causes $\overline{Q}$ to be high. A high value for $\overline{Q}$ turns on the transistor switch, which drives the DISCHARGE pin toward ground. The output stage is an inverting buffer, so SET causes the output to go high, while RESET causes the output to go low.

Begin by connecting a 555 as shown in Fig. 9.6(a) and observe the output. When the output is high, the capacitor is charging through $R_A$ and $R_B$. When the capacitor voltage $V_C$ exceeds $\frac{2}{3}V_{CC}$, the DISCHARGE pin is driven toward ground and the capacitor discharges across $R_B$. The cycle repeats once $V_C$ falls below $\frac{1}{3}V_{CC}$. The frequency is predicted as

$$f = \frac{1}{0.7(R_A + 2R_B)C}.$$

(9.3)

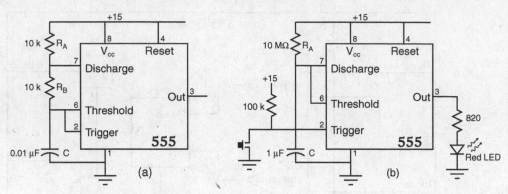

**Fig. 9.6.** (a) 555 timer IC used as an oscillator; (b) 555 timer IC used as a one-shot or timer.

▷ Sketch the output waveform and briefly explain the operation of this circuit. Is the output symmetric? If not, why not?

▷ Derive Eq. 9.3 and compare the measured output frequency with the predicted oscillation frequency.

▷ Examine the voltage $V_C$ across the capacitor. Record its minimum and maximum values. Do they make sense?

▷ Try replacing $R_B$ with a short circuit – what happens? Explain why. Put $R_B$ back for the next part.

▷ Try changing $V_+$ to 5 V and observe how the output changes. To what extent does the output frequency depend on supply voltage?

Now connect a 555 as shown in Fig. 9.6(b). The output should be a 'one-shot' pulse of duration

$$t = 1.1 R_A C. \tag{9.4}$$

The output pulse is triggered by the push-button switch, which causes the TRIGGER input to go to ground. (Note: the output will remain high indefinitely if the TRIGGER input is held at ground, so one should ensure that the trigger pulse is shorter than the desired output pulse!) Time the output pulse by observing the LED.

▷ Briefly explain the operation of this circuit. What prevents this circuit from oscillating?

▷ Measure the output-pulse duration for several values of $R_A$ and $C$. Tabulate your results.

▷ Derive Eq. 9.4. Are your data consistent with this expression? If not, why not?

## 9.2 Additional experiments

### 9.2.1 Alarm!

You can configure the 555 to sound an alarm when prompted by an external signal. The alarm is simply a 555 oscillator with the output connected to a speaker. To prevent the alarm from sounding continuously, ground is applied to the RESET input pin, which overrides the TRIGGER and THRESHOLD pins and forces the output near ground. When used in this way, the RESET line is said to 'enable' the 555, enabling oscillation when high while disabling oscillation when low.

Build the circuit as shown in Fig. 9.7. Use a long wire, which simulates a security loop. Cut or unplug the wire and hear the alarm!

The alarm can also be configured to sound for a specified duration using two 555s by combining the timer from Fig. 9.6(b) with the alarm circuit. Connect the alarm's RESET input to the timer's output line. Trigger the timer and the alarm will sound for the duration of the timer's output signal.

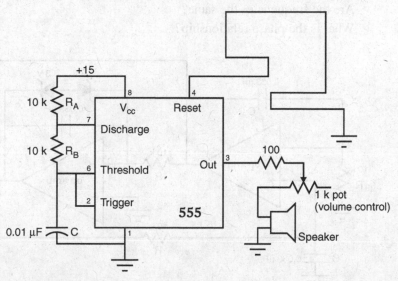

**Fig. 9.7.** 555 timer configured as an alarm.

You can also make a pulsing alarm by using two oscillators. The first should oscillate with a period of a few seconds. The output of this oscillator is then connected to the RESET line of the second oscillator. The second oscillator's output is connected to the speaker and can oscillate with any audio frequency of your choice.

### 9.2.2 Sine/cosine oscillator

Perhaps surprisingly, the sine wave is one of the most difficult waveforms to produce. As you saw when you differentiated the sine output of the PB-503's built-in function generator, the PB-503 uses a piecewise-linear approximation to a sine function.

The circuit of Fig. 9.8 should produce a better approximation. It should oscillate at

$$f = \frac{1}{2\pi RC},\tag{9.5}$$

as long as $R_1 < R$. So you don't expect it to work with $R_1 = 10$ k, unless Eq. 9.5 happens to be satisfied due to resistor manufacturing tolerances. You can vary $R_1$ so as to satisfy Eq. 9.5 by using the 10 k pot for $R_1$ or by adding other resistors in parallel with the 10 k.

Set up the circuit and apply power. Once you have it oscillating, look at both outputs on the dual-trace oscilloscope.

▷ Are the frequencies the same?

▷ What is the phase relationship?

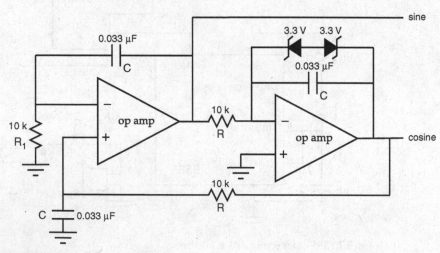

**Fig. 9.8.** Sine/cosine oscillator.

▷ Measure the peak-to-peak voltage of the cosine wave. Is it what you would expect, considering the diodes being employed in the circuit?

Try to explain how the circuit works. What role is played by the two back-to-back Zener diodes?

### 9.2.3 Active bandpass filter

Active filters have significant advantages over passive ones, including
- lower cost due to replacement of expensive inductors by capacitors phase-shifted via feedback;
- high input impedance and low output impedance;
- ease of tuning over a wide frequency range.

They also have significant disadvantages such as frequency response limited by the bandwidth of the op amp and the need to provide power to the op amp. There are a variety of configurations that can be used for low-pass, high-pass, and bandpass applications; for a more extensive discussion, see e.g. Horowitz and Hill, *The Art of Electronics*, or Simpson, *Introductory Electronics for Scientists and Engineers*.

Wire up the circuit of Fig. 9.9, apply power, and drive it with a sine wave of about 1 V peak-to-peak. Vary the driving frequency until you find the maximum output amplitude. What are the center frequency $f_0$ and upper and lower $-3$ dB points $f_H$ and $f_L$ of the frequency response? If you adjust the output voltage to be 14.1 V peak-to-peak at $f = f_0$, you can easily find the $-3$ dB points by varying the frequency until the output voltage is 10.0 V peak-to-peak. Measure and graph the gain *vs.* frequency for a reasonable frequency range.

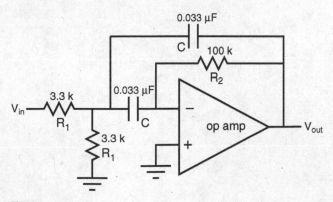

**Fig. 9.9.** Active bandpass filter.

Compare $f_0$ with the theoretical value

$$f_0 = \frac{1}{2\pi C}\sqrt{\frac{2}{R_1 R_2}}, \tag{9.6}$$

and compare the voltage gain at the center frequency with the theoretical value

$$A_0 = -\frac{1}{2}\frac{R_2}{R_1}. \tag{9.7}$$

The 'quality factor' for a bandpass filter is

$$Q = \frac{f_0}{f_H - f_L}. \tag{9.8}$$

Thus, a narrow passband corresponds to high $Q$ and a broad passband to low $Q$. Compare your observed $Q$ to the theoretical value for this filter,

$$Q = \sqrt{\frac{R_2}{2R_1}}. \tag{9.9}$$

Try to explain how the circuit works.

# 10 Combinational logic

In this chapter you will be introduced to digital logic. You will build some logic circuits out of discrete components and some out of integrated circuits, and familiarize yourself with the 7400 series of CMOS (*complementary metal-oxide-semiconductor*) and TTL (*transistor–transistor logic*) integrated circuits and their basic operation.

**Note**

The kinds of things one thinks about in digital logic are almost completely different from those in analog electronics.

**Apparatus required**

Breadboard, oscilloscope, multimeter, 100 $\Omega$, 330 $\Omega$, 1 k, 2.2 k, and 3.3 k $\frac{1}{4}$ W resistors, two VP0610L and two VN0610L MOSFET transistors, three 2N3904 transistors, three diodes, one LED, one red LED (optional), 74HC00, 7432, 7485, 7486 TTL or TTL-compatible logic chips, logic switches, and logic displays.

## 10.1 Digital logic basics

In this section we introduce the 7400 series of CMOS and TTL digital-logic chips. Unlike the analog ICs you've used up to now, which can output any voltage within some range determined by the power-supply voltages, digital-logic ICs employ only two ranges of output voltages, referred to as *logic levels*, about which more below. These levels can be used to represent *true* or *false* logical conditions or the zero and one of binary arithmetic.

The 7400 series is not the only logic series, nor are CMOS and TTL the only types of logic circuitry; however, they are the most commonly used. Other logic families include the CMOS 4000 series and the ECL (*emitter-coupled logic*) 10 000 and 100 000 series. Each logic family has its own logic levels, speed, and recommended supply voltages.

The integrated circuits you will be using now are much more specialized than the general-purpose 741 op amp and 555 timer. They feature much higher bandwidth, with typical transition speeds of order volts/nanosecond (in contrast to the volts/microsecond slew rate of the 741). While greater complexity often means higher cost, the basic chips in the 7400 families (such as the 74HC00, 74LS00, and 74ACT00) cost less than $0.50 each in small quantities, with the more complex chips ranging toward several dollars.

### 10.1.1 Logic levels

Digital chips employ two voltages to represent two possible states. These voltages are called *logic levels* and can be used to represent the two states of Boolean algebra[1] as well as the two digits of binary arithmetic. There are three ways of referring to logic levels:
- true and false,
- zero and one, and
- high and low.

In TTL logic, a voltage exceeding +2 V is called *high*, while a voltage less than +0.8 V is called *low*. To ensure *noise margin*, TTL outputs are guaranteed to put out at least +2.5 V in the high state and at most +0.4 V in the low state (see Fig. 10.1). This means that, even in the presence of up to 400 mV of noise, an output low will be recognized as low by the input of the next logic circuit, and similarly for high. The comparable CMOS levels are +3.5 and +1.5 V for the inputs and +4.5 and +0.5 V for the outputs.

While TTL chips are always powered from a +5 V supply, many CMOS chips are tolerant of supply voltages ranging from +2 to +6 V. It is therefore convenient to reference CMOS logic levels to the power-supply voltage $V_{CC}$. The minimum input voltage interpreted as a CMOS high ($V_{IH}$) equals $0.7 \times V_{CC}$, while the maximum input voltage interpreted as a CMOS low ($V_{IL}$) equals $0.3 \times V_{CC}$. The output voltages $V_{OH}$ and $V_{OL}$ will vary with supply voltage as well.

The 'high/low' nomenclature is unambiguous, since it directly characterizes voltages, but the other two nomenclatures rely on a convention,

---

[1] Also known as symbolic logic.

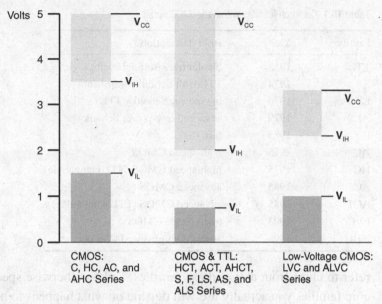

**Fig. 10.1.** Logic levels for various 7400-family lines. $V_{CC}$ is the most positive voltage; $V_{IL}$ and $V_{IH}$ are the maximum input low and minimum input high voltages.

which can be assigned either of two ways. The more common convention is *positive logic*: high=1=true, low=0=false. But there are occasionally situations in which it is more convenient to employ *negative logic*, in which high=0=false and low=1=true. (Another way to think about logic circuits is in terms of *assertion-level* logic, which is a hybrid of positive and negative logic that we will introduce in the next chapter.)

## 10.1.2 Logic families and history

As indicated in Table 10.1, CMOS and TTL chips come in a plethora of types, each with its own speed, power dissipation, input load, and output-current characteristics. This reflects the historical development of the various series. Since our initial purpose is to become familiar with the logic, the details of speed and power are (for now) unimportant, but you will need a general familiarity with them so as not to be in the dark when you encounter them in future. Also there are restrictions on *fanout* (the number of inputs that an output can drive) which matter when actually designing circuits. For example, one LS-TTL output can drive twenty LS-TTL inputs, but only four S-TTL inputs. In what follows we use '7400' generically to

**Table 10.1.** Common families within the 7400 series.

| Family | Year | Brief description |
| --- | --- | --- |
| TTL | 1968 | bipolar transistor–transistor logic |
| S | 1974 | TTL with Schottky transistors |
| LS | 1976 | low-power Schottky TTL |
| ALS | 1979 | advanced low-power Schottky |
| F | 1983 | fast TTL |
| HC | 1975 | high-speed CMOS |
| HCT | 1975 | high-speed CMOS (TTL compatible) |
| AC | 1985 | advanced CMOS |
| ACT | 1985 | advanced CMOS (TTL compatible) |
| LVC | 1993 | low-voltage CMOS |
| AHC | 1996 | advanced high-speed CMOS |

refer to chips from any of these families. Unless otherwise specified, the chip families you actually use will depend on what happens to be on hand in the laboratory or (if you are working through this book on your own) on what you happen to find available.

## Part numbers

The original TTL chips were the 7400 series and the corresponding 'Mil Spec' (military specification) 5400 series; these became popular in the 1970s. TTL chips are labeled with part numbers that begin with a letter code (such as 'SN') that is typically different for each manufacturer (see Fig. 10.2); then comes the '74' that identifies the device as belonging to the 7400 series; then there may be a letter code that identifies the family; then the number that identifies the particular device (e.g. 00 for a quad NAND gate, 01 for quad NAND with *open-collector* outputs, 02 for quad NOR gate, 74 for dual D-type flip-flop, etc.); and finally there may be letters that indicate package style, reliability, degree of testing by the manufacturer, etc. (For example, the MC74LS00ND is a Motorola LS-TTL quad NAND gate in the plastic dual-in-line package with 160 hour 'burn-in' testing.)

## Pinouts and data sheets

Data sheets are available on the web or in data books produced by the chip manufacturer. The sheets specify the function of each pin of the IC package and provide detailed data on chip performance. In general, 7400-series chips have compatible pinouts independent of the family. For example, the

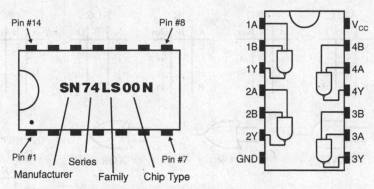

**Fig. 10.2.** Labeling of 7400-series chips.

pinout for a 74HC00 quad NAND IC is the same as the pinout for the 74LS00 quad NAND IC. It is always a good idea to review the data sheet before using any logic chip.

Another consistency is in the pin numbering scheme. If you orient the chip so that the pins are bending away from you and the end that has a notch or a dot is pointing to the left, pin 1 is the one at the lower left of the package. The numbering then proceeds sequentially around the chip in a counterclockwise direction, such that the highest-numbered pin is at the upper left. Almost always, when you orient the chip this way, the writing on the top will be right-side-up (see Fig. 10.2).

### 10.1.3 Logic gates

There are six basic logic gates, as shown in Fig. 10.3. These gates are sufficient to implement all logic functions, although the more complex functions are also available as specialized chips – multiplexers, decoders, etc. (to be discussed later) – which can simplify design as well as reduce cost. Although Fig. 10.3 shows only two-input gates, versions also exist with three, four, or even eight inputs.

Note that NAND and NOR are opposites to AND and OR: NAND equals *not AND* while NOR equals *not OR*. Also note that the NAND is a negative-logic OR while the NOR is a negative-logic AND. (This follows from DeMorgan's theorem, about which more below.) The small circle shown at the outputs of the NAND and NOR is shorthand for an inverter – a NAND is equivalent to an AND followed by an inverter, and similarly for NOR.

| AND | | |
|---|---|---|
| A | B | Out |
| L | L | L |
| L | H | L |
| H | L | L |
| H | H | H |

| NAND | | |
|---|---|---|
| A | B | Out |
| L | L | H |
| L | H | H |
| H | L | H |
| H | H | L |

| OR | | |
|---|---|---|
| A | B | Out |
| L | L | L |
| L | H | H |
| H | L | H |
| H | H | H |

| NOR | | |
|---|---|---|
| A | B | Out |
| L | L | H |
| L | H | L |
| H | L | L |
| H | H | L |

| XOR | | |
|---|---|---|
| A | B | Out |
| L | L | L |
| L | H | H |
| H | L | H |
| H | H | L |

AND          OR          XOR

INVERTER (NOT)

| A | Out |
|---|---|
| L | H |
| H | L |

NAND          NOR          INVERTER

**Fig. 10.3.** Standard logic gates with truth tables.

After the inverter, the NAND is the simplest logic gate to construct from discrete components, and historically it was the most commonly used gate. NAND is a 'universal' logic function, in that the other logic functions can all be created using NAND gates. For example, connecting together the two inputs of a NAND creates a one-input gate – an inverter. Thus, if the output of a NAND is connected to both inputs of a second NAND, the result is equivalent to an AND gate.

### 10.1.4  Summary of Boolean algebra

Logic operations are best described using Boolean algebra. While a detailed exposition of Boolean algebra is beyond the scope of this text, we give here a brief introduction and mention a few useful points. Consider two logical variables $A$ and $B$, which can take on the values true and false. We can then denote logic operations as follows:

- the logical AND of $A$ and $B$ is denoted as $A \cdot B$;
- the logical OR of $A$ and $B$ is denoted as $A + B$;
- the logical XOR of $A$ and $B$ is denoted as $A \oplus B$;
- the inverse of $A$ is $\overline{A}$, and the inverse of $B$ is $\overline{B}$;
- the logical NAND of $A$ and $B$ is denoted as $\overline{A \cdot B}$;
- the logical NOR of $A$ and $B$ is denoted as $\overline{A + B}$.

As in any algebra, the rules of Boolean algebra allow theorems to be derived starting from axioms. An alternative way to prove a theorem in

| A | B | A·B |
|---|---|---|
| L | L | H |
| L | H | H |
| H | L | H |
| H | H | L |

| A̅ | B̅ | A̅+B̅ |
|---|---|---|
| H | H | H |
| H | L | H |
| L | H | H |
| L | L | L |

| A | B | A+B |
|---|---|---|
| L | L | H |
| L | H | L |
| H | L | L |
| H | H | L |

| A̅ | B̅ | A̅·B̅ |
|---|---|---|
| H | H | H |
| H | L | L |
| L | H | L |
| L | L | L |

**Fig. 10.4.** DeMorgan's theorems expressed symbolically.

Boolean algebra is by 'exhaustive demonstration', i.e., to write down the *truth table* – for every possible input state, work out the value of the output and write it down.

Two relationships (known as DeMorgan's theorems) are particularly useful:

$$\overline{A \cdot B} = \overline{A} + \overline{B}. \tag{10.1}$$

and

$$\overline{A + B} = \overline{A} \cdot \overline{B} \tag{10.2}$$

These are illustrated schematically in terms of logic gates as well as with truth tables in Fig. 10.4.

## 10.2 CMOS and TTL compared

### 10.2.1 Diode logic

We begin our consideration of the electronic implementation of logic gates with the simplest example. Since diodes pass current in only one direction, they can be used to perform logic, as shown in Fig. 10.5. Assuming logic levels equal to either 0 or +5 V, it is easy to show that the output voltage is near +5 V (actually one diode drop below +5 V) if both inputs are +5 V, and zero otherwise. If we assume the positive-logic convention (high = true, low = false), this is equivalent to a logic AND of the two inputs. (You may want to verify this by building this circuit on a breadboard and trying it out.)

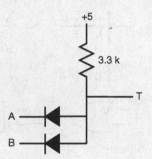

**Fig. 10.5.** Two-input diode gate.

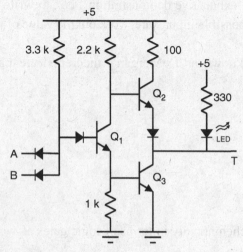

**Fig. 10.6.** Diode–transistor NAND gate using 2N3904s. This resembles the circuitry actually used inside the 7400, except that in the 7400 a two-emitter transistor substitutes for the input diodes. The resistor values are approximate and vary with TTL family. Also shown is a simple LED logic-level indicator being driven by the NAND gate's output.

### 10.2.2 Transistor–transistor logic (TTL)

Using the diode-logic concept as an input stage, TTL logic was developed during the 1960s. A TTL-inspired NAND gate constructed using diodes and bipolar transistors is shown schematically in Fig. 10.6. The operation of the circuit can be analyzed as follows.

- If either input is near ground, transistors $Q_1$ and $Q_3$ will be off. $Q_2$ will consequently be on (saturated), which causes the output (T) to be about two diode drops below $V_{CC}$.
- On the other hand, if both inputs become high, $Q_1$'s base voltage increases, causing $Q_1$ to turn on, which turns on $Q_3$. $Q_2$'s base voltage

drops, turning $Q_2$ off. Since $Q_3$ is then in saturation, the output T is close to ground.

Assuming the positive-logic convention, the output voltage thus represents the logic NAND of the inputs.

The output structure is commonly referred to (fancifully) as a 'totem-pole' output, since it consists of a group of stacked components. Several improvements have been made since the original introduction of TTL logic, resulting in numerous family lines. Many are listed in Table 10.1.

### Quirks of TTL inputs and outputs

The gate circuit of Fig. 10.6 illustrates some peculiarities of TTL inputs and outputs. In general, TTL outputs cannot be relied upon to source even small amounts of current; however, they can sink tens of milliamperes of current. Note how the LED logic indicator has been constructed to take advantage of this – the LED is lit when the output is low! Note also that TTL *inputs* source current while held low, but sink negligible amounts of current while held high.

The input and output current specifications vary among the different family lines. Refer to the data sheets for details.

Note that the LED logic indicators built into the PB-503 operate in positive logic, whereas for TTL, home-built logic indicators (as in Fig. 10.6) operate in negative logic.

### 10.2.3 Complementary MOSFET logic (CMOS)

In the 1970s a new family of logic chips was developed using MOSFETs instead of bipolar transistors. The basic MOSFET logic gate uses *complementary* N-channel and P-channel MOSFETs, and is thus called CMOS. Fairchild first introduced the 4000 series of CMOS logic, but the popularity of TTL led manufacturers to develop TTL-like CMOS families, such as the 74C, 74HC, 74HCT, 74AC, and 74ACT. The families with a 'T' in their part numbers are fully compatible with TTL logic levels, while the ones without a 'T' are logic- and pinout-compatible with TTL but operate with CMOS logic levels. Since the 1990s, CMOS logic ICs have surpassed TTL logic in popularity; however, TTL logic is still manufactured and is quite common.

MOSFET operation resembles that of the JFETs studied previously. As shown schematically in Fig. 10.7, in an N-channel MOSFET, a positive

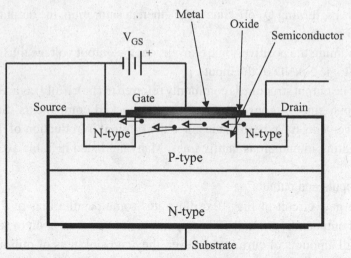

**Fig. 10.7.** Schematic representation of an 'enhancement-mode' N-channel MOSFET.

gate–source voltage attracts current carriers into the channel between the drain and source, allowing current to flow. A gate–source voltage near zero (or negative) closes the channel and prevents current flow between drain and source. When the channel is open, the drain–source resistance is quite small ($\approx 10$–$100\ \Omega$), while the drain–source resistance is large (of order megohms) when the channel is closed. The MOSFET thus acts like a switch.

Notice that to turn on an N-channel MOSFET, the gate voltage is brought *positive* with respect to the source. For a JFET, this would result in a large current flowing through the gate to the source since the gate–source junction would act as a forward-biased diode. This does not occur in MOSFETs since the gate and channel are separated by a layer of insulating oxide. The oxide layer allows the electric field to penetrate without allowing current to pass. The input impedance for MOSFETs is consequently even greater than for JFETs! Also notice that, unlike JFET operation, the drain–source channel is normally closed. A positive gate voltage of a few to several volts is required to open the channel and allow current to flow.

For P-channel MOSFETs, the voltage relationships are reversed: the channel is open when the gate–source voltage is negative and closed when the gate–source voltage is near zero (or positive).

Combining an N-channel and P-channel MOSFET gives a complementary pair of switches that open and close opposite each other. Two

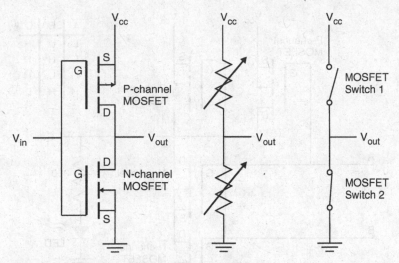

**Fig. 10.8.** Schematic representations of a CMOS inverter constructed using one N-channel and one P-channel MOSFET.

MOSFETs can thus be combined in parallel to create an inverter, as shown in Fig. 10.8. A NAND gate can be constructed by adding two more MOSFETs (see Fig. 10.9). The inverter operates as follows.

- When the input is high, the N-channel MOSFET is on and has a low resistance between drain and source, while the P-channel MOSFET is off and has a high resistance between drain and source. The output is thus connected through a low-resistance path to ground and goes low.
- If the input is near ground, the P-channel MOSFET is on while the N-channel MOSFET is off. The output now has a low-resistance path to $V_{CC}$ and is pulled high.

A drawback is that if the input is at an intermediate voltage, then both channels are partially open, which results in current flowing through both FETs from power to ground. While CMOS logic draws very little power when in a steady state (i.e., all signals either high or low), it thus draws much more power when signals are switching between high and low.

Another drawback of MOSFETs is their sensitivity to static-electricity discharge. The oxide insulating layer between the gate and the channel is usually quite thin and easily damaged. Static charge accumulating on a human in cold, dry conditions can easily develop a potential of several thousand volts. Although the total energy released is small, if this is discharged in a spark to a MOSFET, it is sometimes sufficient to blast a hole

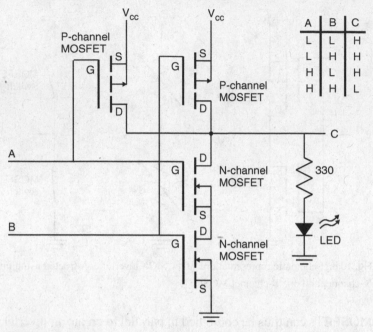

**Fig. 10.9.** Schematic representation of a CMOS NAND gate with LED logic-level indicator.

through the oxide layer, destroying the MOSFET. You've probably heard of static-sensitive electronics such as computer memory cards. These cards are constructed using MOSFETs.

To minimize the accidental destruction of components, we've chosen medium-power MOSFETS for this lab – these are much more robust (and less sensitive to static) than the tiny, high-speed MOSFETS used inside high-density chips.

## 10.2.4 Powering TTL and TTL-compatible integrated circuits

Unlike the 741 and 311 chips that you've used previously, the logic chips to be used here will require only a single +5 V supply rather than separate positive and negative supplies. As in the case of the 741, the actual power connections are often omitted in schematic diagrams, but you must not forget to make those connections. Note that not all chips have the same number of pins; furthermore, even among chips with the same number of pins, not all follow the same convention in power-supply connections. You **must** look up the pinout of each chip you employ!

Be careful! Note that TTL and CMOS ICs are guaranteed to be destroyed if by mistake you power them backwards! They may also be destroyed if you apply a voltage higher than 7 V to any pin (5.5 V for the original 7400 TTL family).

## 10.3 Experiments

### 10.3.1 LED logic indicators and level switches

The PB-503 breadboard has eight built-in LED indicators. These are intended for use as logic-level displays. A high level applied to the input will light the LED – try connecting one to ground or +5 V and see that it works. (If you ever find a need for more than eight indicators, you can augment the built-in indicators with individual LEDs in series with several-hundred-ohm current-limiting resistors; this is also a good solution if your breadboard lacks built-in logic indicators.)

▷ Drive an LED logic indicator with a variable voltage between 0 and 5 V. Explore the threshold voltage and compare with the CMOS and TTL logic levels.

Two single-pole double-throw (SPDT) switches are available at the lower right-hand side of the PB-503. The PB-503 also provides a bank of eight 'level' switches in a unit located near the lower-left corner. These can be used as logic-level switches, as shown in Fig. 10.10. The switches make a connection either to power or to ground. The power voltage depends on the setting of a 'master' switch located to the right of the eight-switch unit, and allows for flexibility when working with low-voltage CMOS ICs or other nonstandard ICs. Older versions of the PB-503 have a bank of SPST DIP

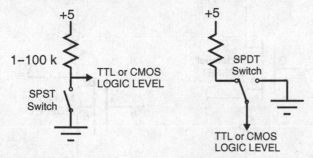

**Fig. 10.10.** Logic-level switch using either an SPST or SPDT switch and a pull-up resistor as shown.

switches instead of logic switches. (If your breadboard doesn't have logic switches, refer to Fig. 10.10 for details on how to use the SPST switch bank as logic-level switches.)

▷ Figure out how to use the logic-level switches to turn an LED indicator on and off. Explain how the indicator works.

CMOS inputs require well defined input voltages; therefore, a CMOS input must always be connected either to power or ground or to a CMOS (or in some cases TTL) output. Unpredictable behavior (as well as excessive current flow from the power supply to ground) will result otherwise. TTL is more forgiving, and SPST switches can easily be used as TTL level switches, even without a pull-up resistor. As we will explore in more detail later, an unconnected TTL input usually behaves as if connected to a logic high: you assert a TTL low logic level at an input by connecting it to ground, whereas an open input acts like a high logic level. With one end of the SPST switch connected to ground, you can thus set the switch 'on' to assert a TTL low input level, or 'off' to assert a TTL high.

### 10.3.2 MOSFETs

To demonstrate MOSFET behavior, construct the circuits shown in Fig. 10.11.

▷ For both N- and P-channel MOSFETs, vary the gate voltage between 0 and 5 V and measure the channel resistance as a function of gate voltage. Note: if you try to measure the resistance directly with the meter, you

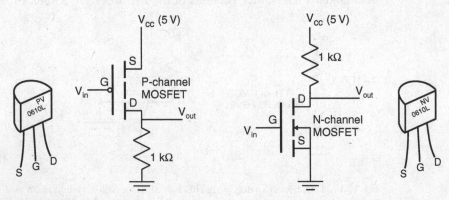

**Fig. 10.11.** Circuits for measuring the channel resistance as a function of gate voltage.

will fail, since the quiescent current flowing in the channel will confuse the ohmmeter! Measure the output voltage and infer the resistance using the voltage-divider equation.

▷ On a single graph, plot the channel resistance for both the N- and P-channel MOSFETs vs. gate voltage. Comment on your results.

As discussed above, the complementary nature of P- and N-channel MOSFETs which you've just demonstrated can be used to create logic gates.

▷ Build the gate shown in Fig. 10.8 and determine its function.

You can easily display the output using an LED indicator. If you use discrete components, don't forget the current-limiting resistor (about 330 Ω) – connect the LED and resistor in series between the CMOS output and ground. In either case, if the LED is on the output is high, and if the LED is off the output is low.

▷ Apply CMOS logic levels to the input and note the output. Make a truth table and verify that the gate inverts the logic level of the input.

▷ The TTL output of the function generator puts out a square waveform whose low voltage level is near zero volts and whose high is near +5 V. Use it to apply a signal to the input, and measure the output *slew rate* (i.e., the transition speed from high to low and from low to high). How does the output transition speed compare with the transition speed of the input square wave?

▷ Measure the input and output impedances and compare your results with your expectations. The output impedance should be measured for both high and low output logic states, using a pull-up resistor to +5 V or a pull-down resistor to ground as appropriate (try a 330 Ω resistor). (If you used a discrete-component LED indicator, you should disconnect it to avoid confusion, or else figure out how to take it into account.)

The inverter can be converted to a NAND gate with the addition of two more MOSFETs. Do so as shown in Fig. 10.9. Connect the output to a logic indicator.

▷ Using logic switches, verify the truth table.

▷ Tie one input to +5 V and the other to a TTL square wave. Measure the transition speed of the output.

▷ Measure the output impedance with both inputs high as well as with both inputs low. Compare these results with the values measured for the inverter. Does the output impedance depend on the output logic level? How does the output voltage level depend on the output load?

### 10.3.3 CMOS NAND gate

From your collection of chips, select a 74HC00 quad NAND gate. Referring to the pinout diagram (Fig. 10.2.), you will note that the ground connection is at pin 7 and +5 V at pin 14, a typical although not universal configuration. Again, **disaster will strike** if you hook up the chip backwards and apply power!

Carefully insert the 74HC00 so that it straddles the central groove of a breadboard section. (When inserting a new chip into the breadboard, you need to pay attention to all of the pins and make sure none of them is bent, since a bent pin will fail to make contact.)

Check that the power is off. Then run jumpers to pins 7 and 14 of the chip from the ground and +5 busses, respectively. For now we are going to use just one of the four gates on this chip, the one whose output is at pin 3 and whose inputs are at pins 1 and 2. Connect pin 3 to an LED logic indicator. Connect pins 1 and 2 to level switches. Ground all other inputs. After double-checking that the connections are correct, turn on the power. Note: the 74HC00 can source only 5 mA of current; therefore, if you use discrete components for your LED logic indicator, you will need to buffer the output through a transistor. This is most simply done using an N-channel MOSFET: connect the source to ground, the gate to pin 3 of the 74HC00, and the LED and current limiting resistor in series from the drain to the +5 V supply.

▷ Connect the inputs to logic switches and try all four input combinations. Record the truth table (in terms of logic levels).

▷ Connect pin 1 to +5 V and drive pin 2 from the TTL output of the function generator. Measure the output rise and fall times and propagation delay of the 74HC00. Compare with the values listed on the 74HC00 data sheet.

For measurements of propagation delay, it is useful to define the transition times of the input and output signals as the times at which they cross 2.5 V, which is about half-way between high and low. A good technique for precise timing of logic signals is to set both scope channels to 1 or 2 V/division and use the scope's vertical-position knobs to put ground 2.5 V below the center of the graticule; then you can easily measure the time at which each signal crosses 2.5 V.

### 10.3.4 Using NANDs to implement other logic functions

As mentioned above, NANDs are a 'universal' logic function in the sense that *any* Boolean-logic function can be constructed from them (the same is true of NOR and XOR). In the early days of logic chips, only NAND was available, since AND and OR functions require more transistors to implement.

▷ Construct the following two-input logic circuits using only NAND gates: AND, OR, NOR.

To figure out the necessary logic, you can make use of DeMorgan's theorems (Eqs. 10.1 and 10.2). Recall that you can use a NAND gate as an inverter if needed – just connect the two inputs together, or tie one input high.

▷ Write down your schematic diagram for each circuit, indicating pin numbers next to each input and output that you use. These should be *logic* diagrams, i.e., gate symbols arranged in a logical order according to their function (like the schematics in the following chapters), *not* chip-level diagrams (such as in Fig. 10.2).

▷ Drive each circuit using two switches on your breadboard and verify that it gives the desired output for each of the four possible input states. Record the truth table for each logic circuit.

Sometimes it is useful to have a gate that outputs a true if one or the other input is true, but not both. Such a gate is called an exclusive-OR (XOR) gate. In Boolean algebra the exclusive-OR of $A$ and $B$ is denoted by $A \oplus B$. As we shall see below, an XOR can be used to test two numbers for equality; it also can be used to produce the sum bit for a 1-bit binary addition.

▷ Construct an XOR circuit out of NANDs. It should turn on an LED indicator when either input is high but not when both are.

It is easy to see how to do this with five gates; it can also be done with only four. Be sure to show, both using Boolean algebra and using truth tables for the intermediate signals in your circuit, that indeed you have implemented an exclusive-OR.

▷ Show how to use an XOR to test two signals for equality. Try it out and show that it works, displaying the XOR output with an LED logic indicator. Does the LED light for equality or for inequality? Why?

## 10.3.5 TTL quad XOR gate

Use TTL chips for this exercise, as floating inputs will not work with CMOS gates.

With a 7486 quad two-input XOR gate, you can test two 4-bit binary numbers for equality.

▷ Using all four XOR gates on the 7486, plus whatever additional gates you need, design a circuit whose output indicates whether the two halves of the 8-bit level switch are set equal (LED on if they are equal, off if they are different).

You will find that this is quite cumbersome using NAND gates, since equality corresponds to all four 7486 outputs being *low*. But it becomes much simpler if you use OR gates such as the 7432.

▷ Record your schematic including pin numbers, build your circuit, and try it out. Allow one or more inputs to *float* by leaving them unconnected. Is your observation consistent with the general property of 7400-series TTL chips that floating input lines default to a high state?

This is a handy feature when testing circuits on a breadboard, but it is good design practice not to rely on it – to be certain of reliable operation and be maximally insensitive to noise, if you are using a gate in a circuit, you should connect all of its inputs either to high or low or to outputs of other gates.

▷ Display the output with a logic indicator and verify that your circuit works for a few representative cases. Record the input and output in each case.

## 10.4  Additional exercises

### 10.4.1  7485 4-bit magnitude comparator

This single chip does the work of the circuit you have just constructed and more. If you have time, look up this chip in a logic data book or on the web, familiarize yourself with its operation, and test it on your breadboard. Describe how the function of this chip differs from the circuit you constructed in section 10.3.5.

# 11 Flip-flops: saving a logic state

Next we turn to *flip-flops*. Also known grandiosely as *bistable multivibrators*, these devices can remember their past. Their behavior thus depends not only on their present input but also on their internal state. Circuits containing flip-flops are termed *sequential* logic circuits, since their state depends on the *sequence* of inputs that is presented to them. A truth table is not sufficient to describe the operation of a sequential circuit – you need a *timing diagram*.

One category of sequential-logic circuits is the *finite-state machine*, which goes through a predetermined sequence of states, advancing to the next state each time it receives a clock pulse. We will encounter some examples of state machines in this chapter, including divide-by-two and divide-by-four counters. A useful tool in understanding a state machine is a *state diagram*, showing the sequence of states through which the circuit passes.

Some of these circuits will probably be the most complicated you have wired up so far. Prepare carefully in advance and you will find you can complete these exercises easily; if you are unprepared, it is likely to take you three to four times as long!

## Apparatus required

The ideas explored in the following exercises apply equally whether CMOS or TTL gates are used – use whichever is most convenient. You will need a quad NAND (7400), a dual D-type flip-flop (7474), and a dual JK flip-flop (74112). In addition, you will need a 74373, a breadboard, two 1 k $\frac{1}{4}$ W resistors, and a two-channel oscilloscope with two attenuating probes. (If possible, avoid unnecessary complications by not mixing CMOS and TTL chips within a single circuit.)

## 11.1  General comments

### 11.1.1  Schematics

For each of the following exercises be sure to write down a *complete* schematic diagram of *each* circuit you build, including pin numbers (but power and ground connections need not be shown). Often when one wires up a circuit off the top of one's head, it fails to work. Writing down the schematic showing all pin numbers is a powerful debugging tool, since it makes incorrect connections much more obvious.

### 11.1.2  Breadboard layout

To keep your breadboard organized, it is a good idea to use one vertical bus for power and another for ground, so that power and ground connections to each chip can be made with short jumpers. A neat layout is easier to debug than a messy one! Remember that the top and bottom busses on the PB-503 are not connected internally.

### 11.1.3  Synchronous logic

The circuits in this chapter introduce the concept of synchronous logic design. In an *asynchronous* circuit, signals can change at any time, requiring the designer to work out in detail the signal propagation delays through all possible paths to make sure that the circuit will work as intended. In contrast, in a synchronous circuit, output signals change only in response to a clock signal. Usually there is one common clock for an entire circuit, and its transitions (low-to-high or high-to-low) – commonly referred to as rising or falling 'clock edges' – are used to synchronize all other state changes in the circuit. This approach leads to circuits that are easier to design and analyze than asynchronous circuits. The synchronous approach is therefore standard in microprocessors (for example) – we have all learned to rate the processing power of computer chips in terms of their clock speed.

### 11.1.4  Timing diagrams

Timing diagrams are essential when analyzing synchronous logic circuits. As shown in Fig. 11.1, timing diagrams for a synchronous circuit usually show the clock signal, one or more inputs, and one or more outputs. Time is shown on the $x$ axis and signal logic levels on the $y$ axis. Signal levels

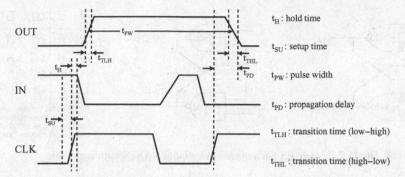

**Fig. 11.1.** Timing diagram with timing definitions for a rising-edge-triggered flip-flop. Note that the rising edge of the clock signal causes the output to change while the falling edge of the clock signal has no effect on the flip-flop.

are shown as either *high* or *low* – we tend not to worry in timing diagrams about analog details such as the exact voltage to which high and low correspond.

Fig. 11.1 defines several important terms. The time between a clock edge and the resulting changing edge of the output is defined as the *propagation delay* ($t_{PD}$). Note how the time is measured from the midpoint between logic low and high. The *setup time* ($t_{SU}$) is the minimum time that an input signal must be stable preceding a clock edge. The *hold time* ($t_H$) specifies the minimum time that the input signal must be stable following a clock edge. The *transition time* measures the time required for a signal to transition from logic low to high ($t_{TLH}$) or from logic high to low ($t_{THL}$).

These timing parameters are specified in the data sheets. Manufacturers guarantee that the IC will perform correctly if the user satisfies the specified minimum values. The specs are usually worst-case values. For example, the measured propagation delay is almost always less than the maximum value specified by the manufacturer. Although manufacturers often specify the typical propagation delay as well, it is *not* guaranteed – safe designers heed the minima and maxima, not the 'typicals'.

## 11.2 Flip-flop basics

### 11.2.1 Simple RS latch

The circuit shown in Fig. 11.2 is the simplest flip-flop: an RS latch made of cross-coupled NANDs (an equally simple RS latch can be made from

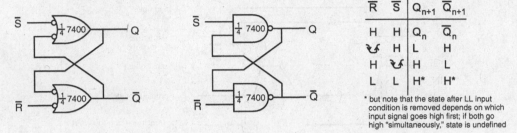

* but note that the state after LL input condition is removed depends on which input signal goes high first; if both go high "simultaneously," state is undefined

**Fig. 11.2.** Simple RS latch made of two-input NANDs with state table.

NORs). R and S refer to RESET and SET (RESET is also known as CLEAR). SET turns Q on while RESET (CLEAR) turns Q off. $\overline{Q}$ is the opposite of Q, except when both S and R are asserted.

## Note on assertion-level logic notation

The inputs in Fig. 11.2 are *active low* – a low input forces a particular output condition, while a high input does nothing. We have therefore taken advantage of DeMorgan's theorem to write the circuit in terms of negative-logic OR gates, rather than blindly using the NAND symbol just because the 7400 is called a NAND gate by its manufacturer. Since the OR gate with inverted inputs has the same truth table as the NAND gate, it is just as good a symbol for $\frac{1}{4}$ of a 7400 as the NAND symbol. And it is actually *better* to use the OR symbol here, since it makes the circuit's operation easier to see at a glance.

The use of DeMorgan's theorem in this way is called *assertion-level* logic notation, and it is a kind of hybrid between positive logic and negative logic – the idea is always to choose the gate symbol that best clarifies the operation of the circuit.

▷ Build and test your flip-flop and record its state table (what its outputs do for each of the four possible input states).

Keep in mind that what the flip-flop does for a given input may depend on past history, so try a few different sequences of input states to make sure you understand what it is doing. In other words, to see the entire state table you need to try each input state for each of the flip-flop's two internal states.

▷ How would the flip-flop's operation be different if it were made of *positive*-logic NORs? (You don't need to build this circuit to answer the question.)

Leave the RS latch in place for use later in this chapter.

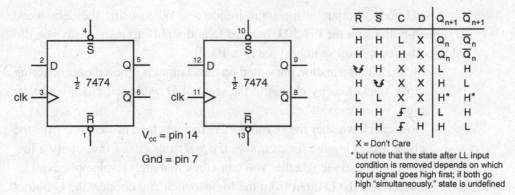

| $\overline{R}$ | $\overline{S}$ | C | D | $Q_{n+1}$ | $\overline{Q}_{n+1}$ |
|---|---|---|---|---|---|
| H | H | L | X | $Q_n$ | $\overline{Q}_n$ |
| H | H | H | X | $Q_n$ | $\overline{Q}_n$ |
| ↻ | H | X | X | L | H |
| H | ↻ | X | X | H | L |
| L | L | X | X | H* | H* |
| H | H | ↵ | L | L | H |
| H | H | ↵ | H | H | L |

X = Don't Care

* but note that the state after LL input condition is removed depends on which input signal goes high first; if both go high "simultaneously," state is undefined

**Fig. 11.3.** 7474 D-type flip-flop with state table.

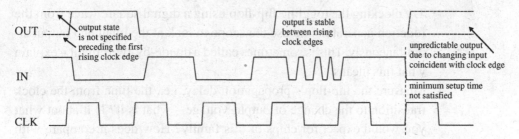

**Fig. 11.4.** Sample timing diagram for a (positive-edge-triggered) 7474 D-type flip-flop.

## 11.2.2 D-type flip-flop

In practice, RS latches are seldom used. The most commonly used flip-flop is the clocked D-type, which remembers the state of its D input at the time of a clock transition, but is insensitive to D at all other times. We will use the 7474 D-type flip-flop (Fig. 11.3), which is sensitive to rising clock edges (low-to-high transitions); in other words, it is *positive-edge-triggered*. (Later in this chapter you will encounter a *negative-edge-triggered* JK flip-flop.)

Note that, in addition to its 'fancy' D and clock inputs, the '74 retains the simple $\overline{\text{RESET}}$ and $\overline{\text{SET}}$ inputs of the RS latch (which are active low since internally the '74 is constructed from NANDs).

▷ Test the $\overline{\text{RESET}}$ and $\overline{\text{SET}}$ inputs and explain what they do; then tie them to +5 V to make sure they are inactive.

Next check out the clocked operation of the '74 (illustrated in Fig. 11.4). Provide the clock signal using a momentary-contact breadboard 'debounced push-button' and the D input using a breadboard logic switch, and display the Q and $\overline{Q}$ outputs using logic indicators. When using the debounced push-buttons on the PB-503, be sure to add a pull-up resistor (as you did for the logic-level switches; see Fig. 10.10).

▷ Show that information presented on the D input is ignored except during clock transitions by changing the state of D while the clock is low or high. What happens?

▷ Now check whether the $\overline{SET}$ and $\overline{RESET}$ inputs take precedence over the clock and D inputs: for example, try asserting $\overline{RESET}$ (i.e. apply a low level to it) and see whether you can clock in a high level applied at D.

▷ Disconnect the D input from the logic switch and connect the $\overline{Q}$ output to the D input to make a toggling flip-flop. (Be sure to deassert $\overline{SET}$ and $\overline{RESET}$.) What happens now when you apply clocks?

▷ Try clocking the toggling flip-flop using a digital square wave from the function generator, and use the scope to look at the input and output simultaneously. This is sometimes called a divide-by-two circuit – explain what this means.

▷ Measure the flip-flop's propagation delay, i.e., the time from the clock transition to the change of output voltage – what is it? Is it about what you would expect for chips of this family? How does it compare with the manufacturer's specifications? Be sure to trigger the scope on Q, not the clock signal.

As mentioned above, a good technique for precise timing of digital signals is to set both scope channels to 1 or 2 V/division and use the scope's vertical-position knobs to overlay both grounds below the center of the graticule; then, you can easily measure the time at which each signal crosses the midpoint between logic low and logic high (1.5 V for TTL and 2.5 V for standard CMOS).

## 11.3 JK flip-flop

The JK flip-flop (Fig. 11.5) is slightly more complicated than the D flip-flop; it can do everything a D can do plus more. The following exercises use a negative-edge-triggered JK flip-flop with SET and RESET. Various chips are available (e.g., the 74HC112 or the 74LS76) – which you use will depend on what you have to hand; however, for the following exercises we recommend using the 74112 JK flip-flop. Be sure to review the data sheet to ensure that you have the correct pinout.

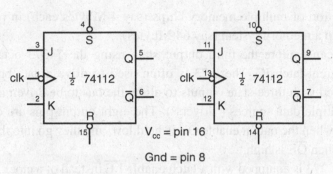

**Fig. 11.5.** Pinout of the 74112 JK flip-flop.

Tie $\overline{\text{SET}}$ and $\overline{\text{RESET}}$ high and explore the JK's clocked operation. (Note that the 74112 senses its J and K inputs only at downward transitions of the clock; hence, it is referred to as negative-edge-triggered.)

▷ Driving the clock from a debounced push-button and J and K from logic switches, check the four possible input states, and write down the JK state table. As in the case of the RS latch above, you need to try each input state for both possible internal states.

▷ Now add an inverter (made from a NAND if you like) from the J to the K input to make a D flip-flop. Drive J from a level switch and write down the state table to verify that the circuit acts like a D flip-flop.

▷ Next remove the inverter and connect J and K together. Try it out and write down the state table. What does this circuit do? How is it different from the toggling D flip-flop?

## 11.4 Tri-state outputs

The ICs that we have used so far have two, and only two, valid output states: high and low. Nowadays, it is common for some logic ICs to have *three* output states.

There are situations in which the designer wishes the output to be neither high nor low! Rather, in such situations, one wants to turn off the output completely, i.e., the output should become a high impedance. The third state is thus referred to as the 'high-$Z$' state. This feature allows multiple outputs to be connected in parallel, as long as all but one of them are in the high-$Z$ state. A common example of the use of tri-state outputs is the connection of multiple memory chips (say 4 Mbytes each) in parallel to build up a memory system (say 64 Mbytes).

We can explore the third output state using the 74373 octal D-type transparent-latch IC. The '373 is often used to drive a data bus, and is equipped with three-state outputs to allow the bus to be driven in parallel by multiple data sources ('drivers'). The eight output pins are driven by the IC when the output enable $\overline{OE}$ is held low, and they go into the high-$Z$ state when $\overline{OE}$ is high.

The '373 is equipped with a latch enable $\overline{LE}$ instead of a clock. Data are transferred from the input to the output while $\overline{LE}$ is high. The outputs are *latched* (held constant) while $\overline{LE}$ is low. This type of flip-flop is called a transparent latch.

▷ Begin by wiring the '373 as shown in Fig. 11.6. (If logic-level switches are available, use them for the $\overline{LE}$, $\overline{OE}$, and D inputs in place of the resistor-SPST switch combinations shown.) Be sure $\overline{LE}$ is high and $\overline{OE}$ is low. Choose any one of the D, Q pairs and apply a valid logic level to the D input. Verify that the Q output follows the D input for both input states. Measure the propagation delay between the input and output.

▷ Set D high and observe the output voltage (voltage at the output pin) as you vary the potentiometer setting (there is no need to be excessively quantitative here). Repeat your observation with D low. Explain why the potentiometer has little effect on the output voltage.

▷ Set $\overline{LE}$ low. Does the output logic level vary as the input changes? Explain your observations.

(a)                                                                 (b)

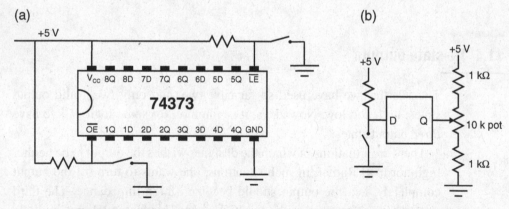

**Fig. 11.6.** (a) Pinout and power connections for the 74373. (b) Input and output connections for testing the tri-state output. Unlabeled resistors merely need to be large enough to prevent excessive current flow from +5 V to ground (e.g. 1 k or larger).

▷ Set $\overline{\text{LE}}$ high and $\overline{\text{OE}}$ high. Observe the output voltage as you vary the potentiometer. Explain your observations. Does the output behave differently for input high and input low?

▷ Now set the input either high or low, record the input state, then set $\overline{\text{LE}}$ low.

▷ Set $\overline{\text{OE}}$ low. The input state at the moment that $\overline{\text{LE}}$ was set low should have been latched internally, independent of the state of $\overline{\text{OE}}$. Does the '373 remember the last input state correctly?

▷ In your own words and based on your observations, explain the operation and features of the 74373.

## 11.5 Flip-flop applications

### 11.5.1 Divide-by-four from JK flip-flops

**Ripple counter**

Cascading two toggling flip-flops makes a divide-by-four circuit, otherwise known as a two-bit counter.

▷ First make the two-bit *asynchronous* or 'ripple' counter shown in Fig. 11.7. (This circuit is asynchronous in that it does not have a common clock signal for all flip-flops.) Clock it from a push-button while looking at the two Q outputs with logic indicators, and verify that it divides by four, i.e., that the output square wave changes state at $\frac{1}{4}$ the frequency of the input clock. Write down its state diagram.

▷ Clock it with a square wave and look at the clock and the outputs on the scope. To see the binary counting pattern, watch clock and $Q_0$, then $Q_0$ and $Q_1$, always triggering on the slower waveform. Write down the

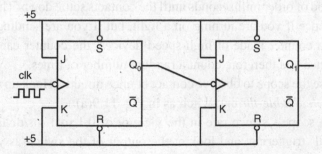

**Fig. 11.7.** Divide-by-four ripple counter.

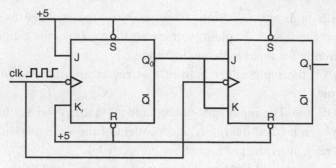

**Fig. 11.8.** Synchronous divide-by-four counter.

timing diagram and label the states with their numeric values (0 through 3), interpreting $Q_0$ as the low-order bit and $Q_1$ as the high-order bit of a 2-bit binary number.

▷ Turn the scope's sweep rate up until you can see the 'ripple': $Q_0$ and $Q_1$ don't change at the same time. Explain why not. (Be sure to look at both edges of $Q_1$.)

### Synchronous counter

Now configure the circuit in its *synchronous* form (Fig. 11.8).

▷ Use the scope to confirm that the ripple is gone, and make a timing diagram showing this. You can see that, unlike D flip-flops, JK flip-flops are natural for building synchronous counters – explain.

Keep your synchronous counter (and RS latch) for the next exercise.

### 11.5.2 Contact bounce

When a mechanical switch closes or opens, there is usually an effect called 'contact bounce'. This means that the switch closes and opens repeatedly for a period of order milliseconds until the contacts settle down. This makes no difference if you are turning on a light, but if you are sending a clock pulse to a counter made of high-speed devices, the counter can react to each bounce and therefore count a random number of times.

First use the scope to observe contact bounce directly. Hook up an SPDT (*single-pole-double-throw*) switch as in Fig. 11.9(a).

▷ Set the scope's sweep rate in the vicinity of 0.1 to 1 ms/division, use 'normal' triggering, and look at the output of the switch as you open and close the contact. Play with the trigger threshold and sweep rate

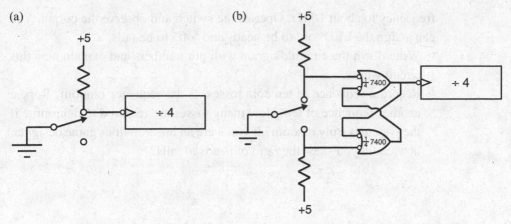

**Fig. 11.9.** (a) Looking at contact bounce by driving a divide-by-four counter from a switch. (b) A NAND latch is used as a debouncer.

to see if you can discern the bounces. Make a sketch of the observed waveform.

▷ Next, clock your two-bit counter from the switch and see what happens. Write down some typical sequences of states. How is the contact bounce affecting the sequence?

### RS latch as debouncer

Now use the RS latch from the first exercise as a switch debouncer (Fig. 11.9(b)). (Although D and JK flip-flops are used for most flip-flop applications, switch debouncing is one area in which RS latches continue to hold their own.)

▷ Connect the switch to the latch as shown, and use the output from the latch to clock the counter. Verify that the counting sequence is now correct. Explain why this works. (The PB-503's momentary-contact switches are already debounced by RS latches built into the unit.)

### 11.5.3 Electronic coin toss

The RS-latch debouncer from the previous section can be combined with a D-type flip-flop to create an electronic coin-toss game. To build the circuit, connect the output of the RS-latch debouncer to the clock input of a D-type flip-flop. Connect the digital output of the function generator to the D input and connect the Q output to a logic indicator. Set the function-generator frequency to about 10 Hz. Operate the switch and observe the output. You can assign the LED 'on' to be heads and 'off' to be tails.

▷ Write down the circuit diagram with pin numbers and explain how this circuit works.

▷ Record a sequence of ten coin tosses. Is the sequence random? Repeat another sequence of ten. How many tosses are required to determine if the system is truly random? If you were an unscrupulous game designer, how could you skew the ratio of heads to tails?

# 12 Monostables, counters, multiplexers, and RAM

This chapter will introduce a variety of techniques that are important in sequential-logic design. Such designs often make use of pulses of various durations. Sometimes a logic pulse of a given width needs to be formed in response to a particular input condition, e.g. to standardize a pulse from a push-button. Monostable multivibrators are the usual solution. In addition to monostables of a given logic family (such as the 74121, '122, '123, etc.), there are also available the family of timer chips (such as the 555); the latter are particularly useful when a long pulse of stable and reproducible width is needed.

In this chapter you will also explore counters and their uses in timing and addressing. As an example of the use of an address counter, you will store and retrieve information in a small memory chip.

Be sure to write down the circuit's schematic, with pin numbers, for every circuit you build. You will find the schematic especially useful should your circuit not work. A simple review of the schematic will often reveal the source of the problem. Futhermore, a schematic is essential when debugging subtle errors.

## Apparatus required

Breadboard, oscilloscope, 7400 NAND, two 7490 and one 7493 counter, 74121 (or similar) one-shot, 74150 multiplexer, 7489, 74189, or 74219 RAM chip, two TIL311 displays, assorted resistors and capacitors.

## Note

The circuits in this lab are rather involved, and many of the details of their design are left for you to work out. You will *need* to work them out *in advance* if you are to have any hope of completing the exercises in a timely fashion!

## 12.1 Multivibrators

Multivibrator circuits fall into three general categories.

* *Astable.* These circuits have no stable state but keep changing from one state to the other, hence the name multivibrator. They are very useful as clocks or oscillators. (You used a 555 as an astable multivibrator in chapter 9.)

* *Bistable.* These circuits can be induced to go from one state to another, and can remain in either state permanently after the input signals have been removed. They are thus stable in both of their allowed states. The more common term for bistable multivibrator is *flip-flop*.

* *Monostable.* These circuits have only a single stable state. They can be forced out of their stable state by a trigger pulse, but they return to it after a very limited period of time. The primary use for monostables is to create pulses of known duration from triggering pulses of shorter, longer, or variable duration. Monostables are widely used to generate 'gate' signals for counter circuits. They are considered to be hybrid analog/digital chips in that the digital output is typically determined by the *RC* time constant of an external analog circuit connected to the chip. Another term for monostable multivibrator is *one-shot*, since it is a device that 'shoots' once (i.e., issues an output pulse) each time it receives an input signal.

Among monostables, the 555 timer is the best choice for pulse widths ranging from milliseconds to hours, and for applications in which the pulse width must be stable to better than 0.05%. In digital designs, monostables such as the 74121 are preferred for pulse widths ranging from about 40 ns to 10 ms, and will operate up to tens of seconds, but their pulse widths are not as predictable or stable.

## 12.2 Counters

In the last lab you wired up a divide-by-four circuit. That was of course a 2-bit binary counter. Counters are so useful that IC manufacturers provide 4-bit (and more) counters as a single chip, with carry-in and carry-out connections that allow them to be 'cascaded' in multiple stages for 8-bit, 12-bit, or greater range. Cascading means connecting multiple chips together (as in the multiple digits of a car's odometer) so that each chip counts when the preceding one 'rolls over' from its maximum count back to zero. Counters are available in both binary and decimal versions and in synchronous and asynchronous ('ripple-through') configurations, with various arrangements of set, reset, and clock inputs.

Four-bit binary counters count from 0–15 and then 'roll over' to 0 again, possibly issuing a carry-out signal to the next stage. (More specifically, synchronous counters issue a carry-out, while for negative-edge-triggered asynchronous counters, the high-order output bit from the preceding stage serves to clock the next stage.)

Decimal counters work basically in the same way as binary counters, except that they roll over (and possibly issue carry-out) at 9 rather than at 15; this makes them useful for driving decimal displays, which are easier for humans to interpret than binary. Note that decimal counters are also referred to as decade or BCD ('binary-coded decimal') counters. Some are actually *bi-quinary* counters, i.e., a divide-by-five stage coupled to a divide-by-two.

## 12.3 Experiments

### 12.3.1 Bi-quinary ripple counter

The 7490 (Fig. 12.1) is a negative-edge-triggered bi-quinary ripple counter. It advances from one state to the next on the falling edge of its clock input. It consists of a 1-bit divide-by-two stage and a 3-bit divide-by-five stage that can be cascaded two different ways. One way produces a decimal counting sequence (0 through 9); the second produces a divide-by-ten sequence in

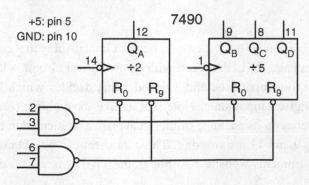

**Fig. 12.1.** Pinout of 7490 decade counter.

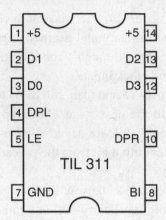

**Fig. 12.2.** Pinout of TIL311 hex display: D0–D3 are data inputs, DPL and DPR connect to LEDs for left and right decimal points, LE = high latches the input data, and BI = high blanks the display.

which the high-order bit is a square wave at one-tenth the frequency of the input clock.

▷ First figure out how to configure a 7490 as a divide-by-ten, clock it with a digital square wave, and verify that the output is indeed a symmetrical (i.e. high half the time and low half the time) square wave at one-tenth the input frequency. Write down the state table and sketch the timing diagram for the four output bits with respect to the clock input. (Also, don't forget to write down your complete schematic with pin numbers.)

▷ Next configure your 7490 as a decimal counter, so that as successive clock pulses are applied it sequences through the states 0–9 in order (0000, 0001, 0010, . . . , 1001).

▷ Display the state of the counter with a TIL311 hexadecimal LED display (Fig. 12.2), as explained in the following paragraphs.

### TIL311 numeric display

The TIL311 is a handy (but expensive) hexadecimal display that combines, in a single package, 22 LEDs, each with its own driver circuit, a latch that can store the four input bits, and a decoder that decides which LEDs to turn on for a given input state. Note that, unlike most chips, the TIL311 has three notches on its package (rather than one), as indicated in Fig 12.2. Also, pins 6, 9 and 11 are missing. The data sheet can be obtained from the Texas Instruments website.[1] Although the TIL311 is a TTL device, it

---

[1]  http://www.ti.com/

will display correctly the output of a CMOS chip, since CMOS logic levels satisfy the TTL input criteria.[2]

To display a 4-bit hexadecimal number, connect the digital signals for the four bits to the pins labeled D0, D1, D2, D3, with D3 being the high-order ($2^3$) bit. Ground BI (blanking input – when high the display is blank) and LE (latch enable – latches input when high). Since you don't want to display a decimal point, leave the DPL (decimal place left) and DPR (decimal place right) pins open. If you wish to experiment with the decimal place LEDs, be sure to use a current-limiting resistor in series with the input pins. See the TIL311 data sheet for additional information concerning these features.

▷ Clock your counter from a debounced switch and confirm that it and the display both work. What are the state table and timing diagram for the four outputs?

▷ Try out the $R_0$ and $R_9$ inputs – what do they do?

▷ Add a second 7490 and TIL311 so that you can count from 0–99. Clock your circuit from a digital square wave at several hertz and verify that it works. Save it for use in the following sections.

### 12.3.2 Monostable multivibrator

The object of this exercise is to design a circuit that generates a pulse of about 500 μs duration. To determine experimentally whether your one-shot is functioning properly, use its output to gate the clock to your two-digit decimal counter, i.e., present the counter with a stream of clock pulses only while the one-shot is firing (see Fig. 12.3). The counter will count up the number of clock pulses, which is proportional to the duration of the pulse from the one-shot.

Fig. 12.4 shows the pinout of the 74121 and 74123 monostable multivibrators. For the $RC$ timing network, use a conveniently sized resistor and capacitor – the timing rules vary by family and type, so be sure to refer to the correct data sheet for your one-shot. In brief, the predicted output pulse width is given by

$$t_w = \begin{cases} \ln 2 R_{ext} C_{ext} & (74121) \\ K R_{ext} C_{ext} & (74LS123) \\ R_{ext} C_{ext} & (74HC123) \end{cases}. \qquad (12.1)$$

---

[2] But the converse is *not* true: TTL logic levels do *not* satisfy the CMOS input criteria.

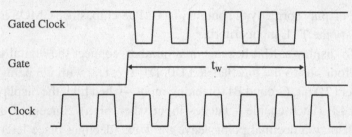

**Fig. 12.3.** Timing diagram for a gated clock signal. Notice how the gated clock signal is simply the logical NAND of the gate and clock signals.

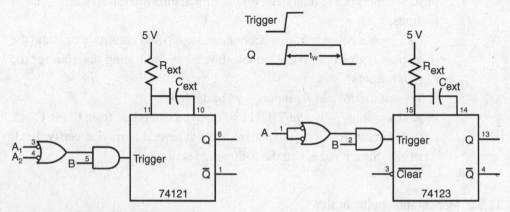

**Fig. 12.4.** Pinout of '121 and '123 one-shots with external $RC$ timing network (see the data sheets for details).

(In the above equation, $K$ is a parameter specifed on the 74LS123 data sheet.)

The output pulse begins following a rising edge at the trigger input. The A and B inputs can be configured either to inhibit triggers, or to produce a trigger from a rising B input or falling A inputs edge. Depending on the specific chip used, there are either one or two A inputs (see Fig. 12.4). There is always only one B input.

▷ Design and build logic that produces a gated clock signal as shown in Fig. 12.3. Be sure to write down the schematic complete with pin numbers.

Use a debounced switch to trigger the monostable. Clock the two-digit counter with the gated clock signal. Choose an appropriate input clock frequency for timing the duration of the one-shot pulse. (Calibrate your time scale by using the scope to measure carefully the period of your digital square-wave clock input.) The clock frequency should be low enough that the counter does not go past 99, but high enough that the width of the one-shot pulse can be measured accurately.

▷ Build the gated counter as described above. Include a push-button reset that zeros the counter. Record the complete circuit diagram including all pin numbers.

▷ Reset the counter and trigger the monostable. What pulse width is implied by the value of the counter, and why? What clock frequency did you choose, and why?

▷ Repeat the measurement about ten times over a period of five minutes in order to determine the reproducibility and stability of the output pulse width. Plot your results as a histogram and compute the mean and r.m.s. (root-mean-square) duration.

(Save your two-digit counter for use in the next section.)

### Note on gating clocks

When gating a clock with a signal that is independent of the clock (e.g. the push-button), a standard problem arises. If the signal from the push-button arrives while a clock pulse is in progress, a pulse of substandard width might be produced (see Fig. 12.5). Similarly, since the one-shot pulse might end during a clock pulse, a substandard pulse might also be produced then. Since substandard clock pulses might fail to meet setup or hold requirements of flip-flops and counters, it is wise to avoid gating clocks whenever possible. When gating a clock is necessary, one normally uses a pulse synchronization circuit such as the 74120 or a pair of cascaded flip-flops to ensure that signals used to gate clocks do not change state during the clock pulse.

▷ Why is your circuit insensitive to this problem?

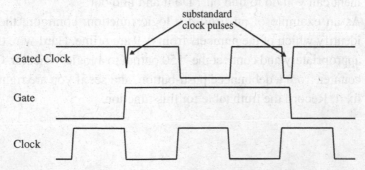

**Fig. 12.5.** Substandard outputs can result when gating clock signals.

The 74123 ICs have several additional features that we haven't explored here. For example, the '123 is a retriggerable monostable equipped with a CLEAR input. The retrigger feature allows the output to persist longer than the time specified by Eq. 12.1 through the application of additional TRIGGER edges while the output pulse is in progress. CLEAR allows the output to be prematurely terminated. See the data sheets for details and operating rules.

### 12.3.3 Multiplexer and finite-state machine

A *multiplexer* (or 'mux') is a device that connects one of $n$ inputs to a single output, under control of an input number in the range 0 to $n - 1$. It can thus be used to select among $n$ different input signals. It can also be used to implement logic functions. For example, by connecting each of the $n$ inputs to low or high in a desired pattern, any desired 1-bit logic function of the input number can be produced.

You can also use a mux plus a counter to generate an arbitrary timing-pulse sequence: on each clock cycle, a different input will be selected, and the output will be either high or low depending on the state of the corresponding input. This is an example of a *finite-state machine* – it repeatedly goes through a cycle of $n$ internal states. Finite-state machines are often useful in control applications (e.g., in deciding when to open the hot-water valve in a washing machine).

Hook up the select inputs (A–D) of a '150 16-to-1 multiplexer (Fig. 12.6) to the outputs of the low-order counter chip from the previous exercise (leave them connected to the hex display also). Note that the '150 has an $\overline{\text{ENABLE}}$ input that needs to be held low. Since the '150 is an inverting mux, if you want its output to be high during counter state $i$, ground data input $i$.

▷ Which select input is high-order and which is low-order? What experiment can you do to find out? Do it and find out.

▷ As an example of an *arbitrary* logic function, configure the '150 to identify which of the numbers from 0–9 are prime. Hard-wire the inputs appropriately and connect the '150 output to a logic indicator. Clock the counter from a debounced push-button, and see if you are right – if not, fix it. Record the truth table for this function.

### 12.3.4 RAM

A *random-access memory* (RAM) is a chip containing a large number of flip-flops, each designated by a unique numeric address. Each flip-flop can

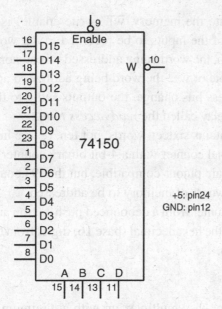

**Fig. 12.6.** Pinout of 74150 16-to-1 multiplexer.

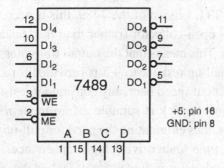

**Fig. 12.7.** Pinout of 7489 16×4 RAM.

be accessed by address for reading or writing. Frequently, the flip-flops are organized into multi-bit words, with each word separately addressable.

For example, the 7489 (Fig. 12.7), 74189, and 74219 are pin-compatible 64-bit RAM chips organized as sixteen words of 4 bits each. Each has four address bits (labeled A through D) for selecting words 0–15, four data inputs ($DI_1$ through $DI_4$) for writing a value into the word being addressed, and four data outputs ($DO_1$ through $DO_4$) for reading the word being addressed. Of course, 64 bits in a chip is nothing nowadays, but it serves conveniently to illustrate the random-access-memory principle using a relatively simple chip.

To write a 4-bit word into the memory, $\overline{\text{WE}}$ (write enable) is brought low. This causes the state of the inputs to be recorded in the word being addressed. When $\overline{\text{WE}}$ is high, the word being addressed is read nondestructively. Regardless of the state of $\overline{\text{WE}}$, the word being addressed appears at the output. When the address bits change, the outputs settle to their new value after a propagation delay called the *read access time*.

Since the RAM chip contains sixteen words, not ten, before hooking it up replace your 7490 decimal counter with a 4-bit binary counter (e.g. the 7493). The 7490 and 7493 are pinout-compatible, but the '93 counts from 0–15, allowing all sixteen words of memory to be addressed.

▷ Clock the 4-bit binary counter from a debounced push-button and verify that it counts through all the hexadecimal (base 16) digits from 0 (binary 0000) to F (binary 1111).

## Open-collector outputs

RAM chips are designed for easy multiplexing with a minimum of additional components, since to increase the total amount of memory available in a circuit one often wants to connect the outputs of multiple RAM chips together. In the case of the TTL version of the 7489, this is accomplished by making the data outputs open-collector (rather than the standard TTL 'totem-pole' output circuit). This means that the output transistors will not operate properly unless a pull-up resistor to +5 is provided for each one. Since we are not worrying about speed here, any convenient resistor in the range of a few hundred ohms to 10 k is suitable. More modern memory chips use three-state outputs, thus eliminating the need for pull-up resistors and also improving the rise time when driving high capacitance.

The 'master enable' ($\overline{\text{ME}}$) signal is provided for use when the outputs of multiple chips are connected together, to allow turning off the outputs of all but one chip. The chip accepts input data and puts out output data only when $\overline{\text{ME}}$ = low. Therefore, be sure to ground $\overline{\text{ME}}$.

▷ Hook up the counter's outputs to the address lines of the RAM and display the output data with your second TIL311. Connect the data inputs and $\overline{\text{WE}}$ to level switches.

▷ Use the address counter and the WRITE-ENABLE switch to program your memory to any desired sequence of hex digits (be sure to record what sequence you choose). If you apply some ingenuity, you can spell out messages using the letters A–F plus I (1) and O (0) (e.g., FEED B0B A D10DE). Then clock the address counter with a digital clock at a frequency of a couple of hertz and watch your message appear! Record the complete circuit diagram with pin numbers and explain how this circuit works.

▷ How could you use the '150 mux to shorten the sequence to any desired fraction of the sixteen addresses? How could you use it to insert blank spaces between words?

# 13 Digital↔analog conversion

In this chapter we will study simple techniques for generating and reading voltage or current levels, i.e., converting between analog (voltage or current) and digital (binary-number) information. The availability of high-speed, easy-to-use, inexpensive digital⇒analog and analog⇒digital converter chips has dramatically changed the way audio and video information are recorded and processed, as well as how computers are used in laboratory research and process control. The process of converting digital information into voltages or currents whose magnitudes are proportional to the digitally encoded numbers is called digital-to-analog (D/A) conversion. The reverse process is called analog-to-digital (A/D) conversion. The devices that carry out these conversions are called DACs and ADCs, respectively.

In this chapter, after building a simple DAC from a digital counter and an op amp, you will continue your exploration of analog/digital conversion by building a 4-bit tracking ADC. Having learned the basic operating principles, you'll use an ADC080x 8-bit successive-approximation A/D chip to *digitize* (i.e., convert to digital) an arbitrary AC signal. The original signal will then be re-created from the digitized data using a DAC080x D/A chip. This exercise will also allow you to explore the limitations of ADC and DAC operations.

Please be sure to work through these circuits in advance, otherwise it is highly unlikely that you will successfully complete the exercises in a timely fashion! Carefully study the manufacturer's data sheets which provide extensive details on operation and performance. As always, complete schematic diagrams significantly improve debugging efficiency.

## Apparatus required

Breadboard, oscilloscope, 74191, TIL311, 311 comparator, 741 op amp, resistors, capacitors, DAC0806 (or similar), ADC0804 (or similar), 7400, 7432, four 7474, 74112, 74138.

## 13.1 A simple D/A converter fabricated from familiar chips

Recall that when an op amp is set up as an inverting amplifier, the non-inverting input is grounded, and the inverting input, which is tied to the output through a feedback resistor, acts as a 'virtual ground'. If a resistor $R$ is connected from a voltage $V$ to the inverting input of the op amp, a current $V/R$ will flow. If you double the resistance, half as much current will flow. Suppose you have four resistors with the resistances $R, 2R, 4R$, and $8R$. The corresponding current flows will be in the proportion 8:4:2:1 (see Fig. 13.1(a)).

A 74191 counter has four outputs $Q_3$ through $Q_0$, with $Q_3$ the MSB (most significant bit) and $Q_0$ the LSB (least significant). In addition it has four parallel-load inputs, count-enable and count-direction (up/down) inputs, and ripple-clock and terminal-count outputs for use when cascading multiple stages. If we feed the counter outputs to the inverting input of an op amp through resistors $R, 2R, 4R$, and $8R$ (in order from MSB to LSB), we get a 4-bit digital-to-analog converter. The current into the feedback resistor will be proportional to the number that corresponds to the state of the counter. Given a suitable feedback resistor such that the op amp does not saturate, the output voltage will be proportional to this current. To produce a desired output voltage, we can load into the counter any desired value; we can also increment or decrement the counter to get a voltage that changes in time in stepwise fashion (see Fig. 13.1(b)). This output can, of course, be observed on an oscilloscope or other measurement device.

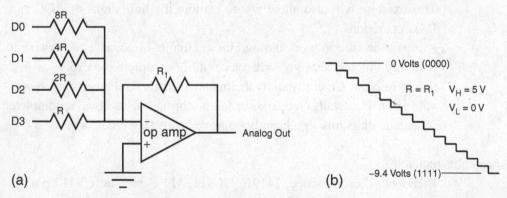

**Fig. 13.1.** (a) Simple D/A converter; (b) output waveform resulting from input counting sequence.

To demonstrate D/A conversion, you will build such a 4-bit DAC. To reduce the chances of hooking up the circuit incorrectly,

▷ Begin by setting up a 74191 counter and make sure it is working properly: hook up its outputs to a TIL311 display, clock it from a debounced switch, and verify that it goes through all sixteen states in order. Test it counting both up and down – you can control which way it counts using the '$\overline{U}$/D' input.

▷ Next, hook up the counter outputs to the summing junction of the op amp through resistors, as described above. Use a 2.2 k resistor to connect to $Q_3$, a 4.7 k resistor for $Q_2$, a 10 k resistor for $Q_1$, and a 22 k resistor for $Q_0$, and connect a 3.3 k feedback resistor. Connect a 1 kHz digital signal to the clock input of the counter and view the analog output on an oscilloscope.

Of course, if we wanted to produce accurate analog output voltages, we would need precision resistors, for example 2.50 k, 5.0 k, 10 k, and 20 k. Moreover, we would need to take into account the inevitable small differences among the high and low levels of the counter outputs. We shall not worry about these refinements, since it is our intention here merely to illustrate the basic idea of D/A conversion.

The output should be a fifteen-step staircase waveform (Fig. 13.1b), with each step having approximately the same height. To see a stable display of the waveform, you can trigger the scope using the falling edge of the MSB.

▷ What full-scale output voltage do you expect (i.e. when the counter is at 15 in decimal or 1111 in binary)? What do you observe?

▷ Is the staircase rising or falling? Why is this? What simple change can you make to reverse the direction of the staircase?

▷ What are the output voltages corresponding to states 4, 5, 6, 7, and 8 of the counter? Measure the four resistances and the high and low voltage levels of the four counter outputs ($Q_0$–$Q_3$), and explain each DAC output voltage.

▷ Write down a complete circuit diagram with pin numbers. Explain in your own words how this circuit works.

Despite the common misconception that modern electronics is strictly digital, analog electronics is still going strong. For all practical purposes, our everyday world is analog. The digital representation of *any* waveform (music, for example) is only an approximation. To smooth out the discontinuities of digitized waveforms requires analog electronics.

## 13.2 Tracking ADC

To measure an analog signal you need to invert the process of D/A conversion. There are various ways of doing this, but, just as division is harder and slower than multiplication, and taking the square-root harder and slower than squaring, analog-to-digital conversion is harder and often slower than digital-to-analog.

Given a DAC, a counter, and a comparator, a simple approach is to increment the counter (starting from zero) until the DAC output crosses the analog input. Using the comparator to compare the analog input to the DAC output, you stop counting when the comparator output switches states. At that point, the counter holds a digital approximation to the magnitude of the input.

A simple variant of this circuit will follow (or 'track') changes in the input voltage. You can turn your 4-bit counter/DAC into such a tracking ADC by driving $\overline{U}/D$ from a comparator that compares the DAC output with the analog input voltage.

Use a potentiometer to make the analog input voltage: connect one end to ground and the other to −15 V. The slider controls the input voltage, which you can vary between 0 V and −15 V. To stabilize the operation of the circuit, use some hysteresis by connecting a series 10 k resistor between the input voltage and the comparator noninverting input and 1 M between the comparator output and the noninverting input (see Fig. 13.2).

▷ Clock the counter at a few hertz and observe its state with the TIL311 as you vary the input voltage. What do you observe?

▷ How should the comparator inputs be configured: which signal should go to the inverting and which to the noninverting input? Is the output number 'homing in' on the expected value? If not, did you perhaps connect the comparator backwards? Explain, and if you did it wrong the first time, fix it.

▷ Record the output numbers for a few different input voltages.

▷ Why is the output number never stable? How (if at all) does this affect the precision of the voltage measurement?

Note that the tracking ADC is slow at following large input-voltage changes, since it has to count through all the intermediate values, but it has good performance if the input voltage changes gradually.

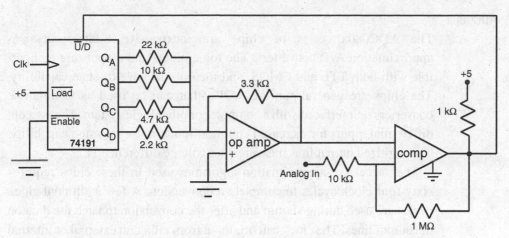

**Fig. 13.2.** Simple A/D converter. (The polarity of the comparator inputs is left as an exercise for the reader.)

## 13.3 080x ADC and DAC chips

### 13.3.1 Successive-approximation ADC

The technique just described is comparable to looking for a word in the dictionary by looking at each word one after the other until you find the right one (or, if the word you're looking for isn't in the dictionary, until you find a word beyond the one you're looking for). As we've seen, this is a fine algorithm for tracking a slowly changing signal, but if the DAC is far from the signal voltage it takes a long time to home in.

A faster approach (and one used in many ADC chips) is based on the binary (or logarithmic) search algorithm, in which at each step you reduce the search range by half. For example, if you're looking up a word in an $n$-word dictionary, first look at word $n/2$. If the word you want comes later in the alphabet, next try word $3n/4$; if your word comes earlier, next try $n/4$. At the next step there are four possibilities: word $n/8$, $3n/8$, $5n/8$, or $7n/8$ – and so on. This method will find any word in at most $\log_2 n$ steps.

In the case of analog-to-digital conversion, the logarithmic search has another name: successive approximation. If there are $n$ output bits, there are $2^n$ possible output values, but instead of trying each *value* in succession, you try each *bit* in turn, starting from the MSB: first generate a 1-bit approximation to the value, then correct it to 2-bit accuracy, then 3-bit, and so on.

## ADC080x

The ADC080x series of chips are inexpensive 8-bit successive-approximation A/D converters. The logic inputs and outputs are compatible with both TTL and CMOS, and the outputs have tri-state capability. The chips are general-purpose ADCs that can be used as stand-alone converters or interfaced with a computer or other logic system. They accept differential inputs for increased common-mode-noise rejection capability. The digitized output thus measures the voltage difference $V_{in+} - V_{in-}$.

The successive-approximation algorithm used in these chips requires sixty-four clock cycles to complete a conversion. A few additional clock cycles are used during startup and after the conversion to latch the data on the output lines. The clock can originate from either an external or internal ('self-clocking') source. The self-clocking option uses an on-chip oscillator (with Schmitt-trigger timing input), in combination with an external resistor and capacitor that determine the period, as shown in Fig. 13.3.

All of the input and output control signals are active-low. There are three input control lines, labeled $\overline{CS}$, $\overline{RD}$, and $\overline{WR}$. An A/D conversion is started

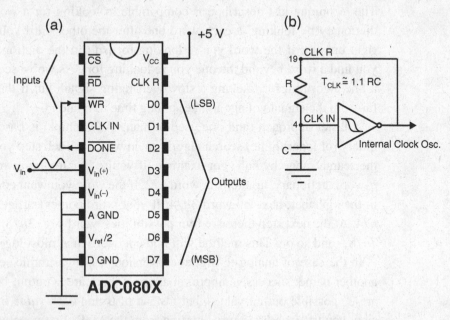

**Fig. 13.3.** (a) Pinout for the ADC080x series of A/D converters. (b) The on-chip self-clocking configuration. Note the locations of the most significant bit (MSB) and the least significant bit (LSB). The 'x' in ADC080x means that multiple versions of this IC exist (e.g. ADC0804).

by bringing $\overline{\text{RD}}$ and $\overline{\text{WR}}$ low simultaneously. $\overline{\text{CS}}$ is used in microprocessor-based applications – for this exercise it should be connected to ground. $\overline{\text{RD}}$ is equivalent to output-enable – $\overline{\text{RD}}$ high puts the outputs into their high-impedance state; when low, the output lines are active – therefore, connect $\overline{\text{RD}}$ to ground. With $\overline{\text{CS}}$ tied to ground, $\overline{\text{WR}}$ simplifies to A/D START (i.e., start the digitization **now**!). Once the digitization completes, the result is latched onto the output bus and a DONE pulse is issued (pin 5).

Connecting $\overline{\text{DONE}}$ to $\overline{\text{WR}}$ (as shown in Fig. 13.3(a)) puts the ADC into a 'free-run' mode, in which the completion of one conversion initiates the next. Once triggered, the ADC will repeatedly digitize the input voltage difference. To ensure free-run status following power-up, $\overline{\text{WR}}$ may need to be brought low manually, so connect $\overline{\text{WR}}$ to a debounced push-button as well as to $\overline{\text{DONE}}$.

The chip has two grounds, called analog ground (A GND) and digital ground (D GND). In many applications, the analog and digital grounds are kept separate to reduce the analog noise introduced by rapidly changing digital signals. We will ignore these effects and use a common ground for both pin 8 and pin 10.

The range for the input voltage difference is determined by the input voltage reference – we are using $V_{\text{CC}}$. A digital output of 00000000 should correspond to an input very close to zero volts, while the output 11111111 should correspond to a voltage near $V_{\text{CC}}$. We shall use single-ended inputs, so you should ground $V_{\text{in}-}$.

(A more complete explanation of the ADC080x's features can be found in the manufacturer's data sheets.)

▷ Connect an ADC080x chip as shown in Fig. 13.3(a). Connect pins 1, 2, 7, 8, and 10 to ground. Connect pin 4 to ground using a 50 pF capacitor. Place a 10 k resistor between pins 4 and 19. Connect pins 3 and 5, and add a connection to a push-button. Using the 1 k pot, apply a variable voltage between +5 V and ground to pin 6. Connect pins 11–18 to LED logic indicators. Connect pin 20 to +5 V power, and leave pin 9 unconnected.

▷ Adjust the input voltage and observe the digital output. What measured input voltage corresponds to the binary output 00000001? What measured input voltage corresponds to the binary output 11111110? Measure several other input-voltage values and plot the input voltage versus digital output. Is the plot linear? What is the input range, and how does the ADC respond to small excursions outside this range (not less than ground and not more than $V_{\text{CC}}$)? With what precision does the ADC measure the input voltage?

▷ Estimate the conversion time by looking at pin 5 with the oscilloscope. (Conversion time is the amount of time an ADC requires to digitize an input voltage.) What is the sampling rate? Does the conversion time depend on the input voltage? Taking into consideration the value of the external $RC$ network, does the measured sampling rate agree with your expectations? If not, why not? (Hints: parallel capacitors add linearly, and some small stray capacitance is common on breadboards such as the PB-503.)

▷ Observe, sketch, and explain the waveforms at pins 4 and 19. Replace the 50 pF capacitor with a 100 pF capacitor. Explain what happens to the waveforms. What happens if you remove the 100 pF capacitor completely (i.e., do not use any external capacitor)? Estimate the value of the stray capacitance. Replace the 50 pF capacitor.

▷ Write a complete circuit diagram with pin numbers, and explain in your own words how this circuit works. Comment on your observations and measurements.

Save your ADC circuit for use in the next part.

## The DAC080x D/A converters

The DAC080x is a family of popular D/A converter chips, of which the DAC0806 is the least expensive – it is an 8-bit DAC with 6-bit precision, while the related DAC0807 and DAC0808 have 7- and 8-bit precision, respectively. Our home-brew D/A converter above used 4 bits to distinguish sixteen different voltage levels. With the DAC0806 we can distinguish not sixteen but sixty-four different voltages (at least). If you need more accuracy, you can use the DAC0807, the DAC0808, or one of the more sophisticated chips that are available. (You may use whichever of these chips is available for this exercise.)

Chips of the DAC080x family output a current that is proportional to the value of the digital input. Some D/A converter chips (e.g., the expensive, and less readily available, NE5018) include an op amp on the same chip. The DAC080x family does not! To convert the DAC's output current to an analog voltage, you can use an external op amp, as shown in Fig. 13.4.

What exactly is the difference between the 'top-of-the-line' DAC0808 and the 'budget' DAC0806? Within the DAC0808 chip are eight resistors of nominal resistance $R, 2R, 4R, 8R, 16R, 32R, 64R$, and $128R$. For 8-bit accuracy, these need to have better than 0.5% tolerance. The DAC0806

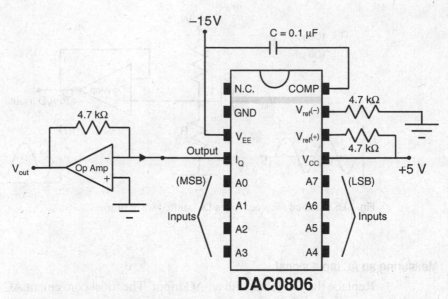

**DAC0806**

**Fig. 13.4.** Pinout for the DAC080x series of D/A chips (with an output op amp added). Note that the bit order is the reverse of that used for the ADC080x, such that A0 is the MSB and A7 the LSB.

also has eight resistors, but their tolerances are not as good, such that the DAC0806 is not guaranteed to put out 256 distinguishable voltage levels, but only sixty-four.

These chips also contain circuitry to buffer the input signals and standardize their voltage levels, to avoid inaccuracies due to voltage variations in logic levels on different input lines. Consult the manufacturer's data sheet to see what else your D/A converter chip contains.

▷ Wire up a DAC080x chip as shown in Fig. 13.4. This configuration will give an output voltage between 0 and +5 V. Connect the outputs of the ADC you built above to the inputs of the DAC.

▷ What is the output voltage that corresponds to the DAC digital input 00000001? What is the output voltage that corresponds to the digital input 11111110? Take a few more data points and plot output voltage versus digital input. Comment on your results.

▷ Measure the precision of your DAC. Does it match the specified precision?

▷ Be sure to write down a complete circuit diagram with pin numbers. Explain in your own words how this circuit works. Comment on your observations and measurements.

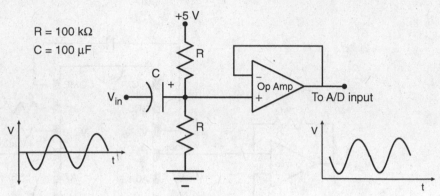

**Fig. 13.5.** Method for producing a DC-shifted waveform.

## Measuring an AC input signal

Replace the DC input with an AC input. The most convenient AC input is a waveform from your function generator. However, the function-generator output is symmetric about ground. The waveform will thus be outside the ADC range 50% of the time! You can DC-shift the waveform as shown in Fig. 13.5. **Warning:** be careful not to exceed the ADC input voltage range! Input voltages larger than $V_{CC}$ or lower than ground may damage the ADC chip.

▷ Explain how the DC shift occurs and the significance of the component values chosen. Suggest at least one other method for DC-shifting the input waveform.

▷ Apply a sine wave of 2 V amplitude centered at +2.5 V to the A/D input. Set the frequency to ~0.5 Hz and try to measure the amplitude using the LED indicators. You can 'freeze' the A/D output using the push-button connected to pin 3. Explain how this works!

▷ Compare the input and output waveforms. How well do they agree? Comment on your observations.

▷ Increase the input frequency (you may need to adjust the amplitude to stay within the range of the A/D input – why?). What happens to the output waveform as the input frequency approaches the sampling frequency? What happens when the input frequency exceeds the sampling frequency? Make several sketches of the input and output waveforms at various frequencies, and comment on your observations. Explain the relationship between sampling frequency and output frequency response.

▷ Experiment with triangle and square waves of various frequencies. Record and sketch a few input and output waveforms, and comment on your results. Be careful not to exceed the allowed input range.

As discussed above, A/D and D/A conversions are merely approximations. Higher precision and higher sampling rates improve the approximation at the expense of increased cost and data size. For example, let's say we're recording the sound track for a TV commercial. To digitize and record a 30 s waveform with 8-bit precision and 10 kHz sampling rate requires 2.4 Mbit of memory storage. Increasing the precision to 16-bit and sampling rate to 100 kHz increases the required space by a factor of 20, to 48 Mbit.

## 13.4 Additional exercises

### 13.4.1 Digital recording

You can convert your A/D–D/A circuit into a digital recording and audio-processing system by adding an audio input, a memory to store the digital data, and an audio output. You can use speakers for both input and output, buffered with suitable op amp circuits. The 32 k × 8 CY62256 memory chips require a 15-bit address counter, which you can make from 7493s or 74191s. What other control circuitry do you need? By varying the clock speed you can trade off fidelity (sampling rate) for message length – e.g. at 1 kHz sampling rate, you would be able to store 32 s worth of sound. Try recording a sound sample, then playing it back at various speeds. You can also try adding nonlinear gain elements (say a logarithmic amplifier) or filtering to see how the sound is affected. Can you figure out a way to program reverb?

For faithful recording, it is important that the input voltage be constant during the entire time of the conversion. This can be accomplished using a sample-and-hold amplifier (SHA) to sample the input and hold it until the conversion is complete. The National Semiconductor LF398 is an SHA IC. Pin 1 is connected to +15 V and pin 4 is connected to −15 V. There are separate analog and logic inputs, on pins 3 and 8, respectively, and the output is at pin 5. Ground pin 7. An external capacitor (1000 pF) is connected between pin 6 and ground. This capacitor is used to hold the analog data until the ADC has had an opportunity to process it.

If you choose to add a sample-and-hold, what additional control logic is needed? Can you hear the difference it makes in fidelity? How would you describe the difference, and how would you explain it?

### 13.4.2 Successive-approximation ADC built from components

To see first-hand how the successive-approximation algorithm works, you can build an 8-bit ADC using TTL or CMOS parts plus a DAC and comparator.

In our successive-approximation circuit (Fig. 13.7) the eight bits are stored in four '74 dual D-type flip-flop chips. The successive-approximation algorithm consists of trying each bit in both the 0 and 1 states, starting from the MSB and ending with the LSB. Each bit starts out at 0 and is then set to 1. If the DAC output exceeds the analog input, the bit is set back to 0, otherwise it is left as a 1. To accomplish this, the circuit goes through a sixteen-state cycle, in order both to set and (possibly) reset each of the eight bits.

The sixteen-state cycle is provided using a '191 four-bit binary counter. The '138 1-of-8 decoder routes clock pulses (how many?) to each flip-flop in turn. The 311 comparator compares the DAC output with the analog input, and its output is connected to the D input of each flip-flop. Note that the DAC080x is a current-sinking DAC, and its output thus becomes more negative as its digital input increases from 0 to 255. The connection shown thus provides negative feedback: if the DAC output is too negative, the currently addressed bit is set to 0 in order to raise the DAC output voltage, and if the DAC output is too positive, the bit is set to 1 in order to lower it.

### Timing and control logic

If you choose to undertake this exercise, you should first spend some time understanding the timing cycle and drawing a complete timing diagram for the control logic (shown in Fig. 13.6 and the lower left-hand corner of Fig. 13.7). Pay particular attention to the sequence and timing of signals at the beginning and end of the conversion cycle, and see if you can figure out the answers to the following questions. In what state does the counter start out? In what state does it end? Exactly what function does each decoder enable perform? (Hint: each performs a slightly different function.) What happens if you attempt to start a new conversion cycle while one is already in progress? Why is it important for the control flip-flops to be negative-edge

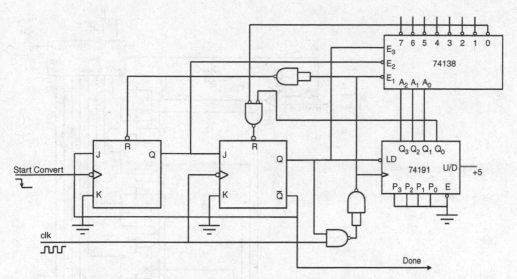

**Fig. 13.6.** Control logic for 8-bit successive-approximation ADC.

triggered: can the clock signal to the '191 glitch (i.e. have a pulse of sub-standard width – see discussion in section 12.3.2)? Why, or why not?

Begin by building and debugging the control logic by itself – but be sure to leave room for the additional chips! (If you prefer, you may choose to build and analyze the simpler control logic described below.) Try to arrange the chips and connecting wires neatly, so that it is easy to see where each wire goes – some color-coding could be helpful. In case a chip needs to be replaced, try to avoid overly tight wiring across the top of any chip.

Clock your control circuit with a digital square wave from the function generator, and provide the 'START CONVERT' signal with a debounced push-button. You should be able to see the full timing sequence on the oscilloscope. Trigger the scope on the output of the second flip-flop and observe each of the other signals as you repeatedly issue 'START CONVERT.' Verify that your timing diagram is correct.

## Complete ADC circuit

Now add the rest of the circuit. Use the breadboard logic indicators to display the data bits. Attach the $\overline{\text{RESET}}$ input to the other debounced push-button.

Measure the DC analog input voltage $V_{\text{in}}$ with a digital voltmeter. Try several voltages over the full range and make a graph of your results.

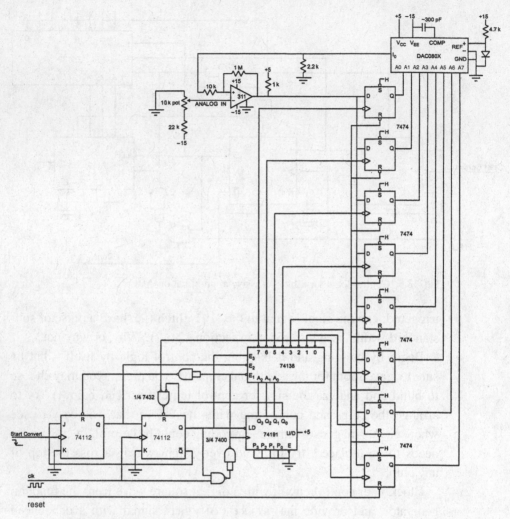

**Fig. 13.7.** 8-bit successive-approximation ADC.

How good is your ADC? Is it linear? What is its zero offset (the number it puts out for $V_{\text{in}} = 0$), and what is its slope constant (volts per output count)? What is its least count (the voltage change corresponding to one ADC count)? What is its full-scale voltage?

Try raising the clock frequency. At what frequency does it stop working? Does this make sense? What do you think limits the conversion speed of this circuit? (Illustrate your answer with the relevant timing diagram.)

Will this ADC always work correctly the first time after power is turned on? Why, or why not?

## Simpler version of control logic

You might want to build and analyze a simpler version of the control logic using parallel-output shift registers instead of the counter/decoder version discussed above. The idea is to produce eight bits, of which seven are 1s and one is a 0, and circulate them around an 8-bit shift register. Connect each shift-register output both to clock and to $\overline{\text{SET}}$ of a flip-flop, so that each flip-flop is first set, and then reset, by the clock edge if the analog output is too big. This simple approach might or might not work, depending on the internal timing of the flip-flop (clock and $\overline{\text{SET}}$ are changing simultaneously) – the question is whether the SET signal goes away within the flip-flop soon enough so as not to override the clock edge. If it does not, you can use the 'diode trick' to delay the clock relative to $\overline{\text{SET}}$. (The diode trick consists of adding a series diode to shift a digital signal in voltage, thereby changing the time at which it crosses threshold.)

# Further reading

The following table indicates the sections or chapters in four popular textbooks where you can find additional background information for each chapter of our text.

| Expt | D. & H. | Barnaal | H. & H. | Simpson |
|------|---------|---------|---------|---------|
| 1 | 1, 6 | A1, A2 | 1 thru 1.02, 1.09, 1.11, 1.32, 12.01 | 1, App. C |
| 2 | 2, 3 | A1 | 1.02–03, 1.07–09, 1.12–16, 1.18–20 | 2, 3, App. A |
| 3 | 4, 5 | A1, A4 | 1.17, 1.25–28 | 4 |
| 4 | 8 | A8 | 2 thru 2.13, 2.15–16 | 5 |
| 5 | 8 | A8 | 3 thru 3.09 | 6 |
| 6 | 8 | A4, A5, A8 | 2.18, 2.14 | |
| 7 | 9 | A6 | 4 thru 4.09, 4.11–12 | 9, 10 |
| 8 | 9 | A6 | 4.09–10, 4.14, 4.19–20 | 10 |
| 9 | 10 | A7, A1 Suppl. | 4.23–24, (5) | 10, 11 |
| 10 | 11 | D1, D2 | 1.10, 8 thru 8.06, 8.08–10 | 12 |
| 11 | 12 | D3, D4 | 8.07, 8.16–17 | 13 |
| 12 | 12, 13 | D5 | 8.14, 8.18–26, 8.34 | 13 |
| 13 | 14 | D6 | 4.16, 9.15–16, 9.20 | 15 |

Key:

**D. & H.:** A. James Diefenderfer and Brian E. Holton, *Principles of Electronic Instrumentation* (Saunders, 1994);

**Barnaal:** Dennis Barnaal, *Analog Electronics for Scientific Application* and *Digital Electronics for Scientific Application* (reissued by Waveland Press, 1989);

**H. & H.:** Paul Horowitz and Winfield Hill, *The Art of Electronics* (2nd edition, Cambridge University Press, 1989);

**Simpson:** Robert E. Simpson, *Introductory Electronics for Scientists and Engineers* (2nd edition, Prentice-Hall, 1987).

# Appendix A
## Equipment and supplies

**General equipment you will need:**

one Global Specialties PB-503 powered breadboard, or equivalent;

one Tektronix TDS 210 Dual Trace Oscilloscope, or similar;

two oscilloscope probes with 10X attenuation;

one digital multimeter with probes;

one power tranformer (12.6 V each side of center tap);

four 50–100 cm banana leads (two red and two black).

## Analog components

| Resistors, $\frac{1}{4}$ W | Number required | Capacitors | Number required |
|---|---|---|---|
| 33 Ω | 1 | 50 pF | 1 |
| 68 Ω | 1 | 100 pF | 1 |
| 100 Ω | 3 | 300 pF | 1 |
| 330 Ω | 2 | 0.0047 μF | 1 |
| 560 Ω | 1 | 0.01 μF | 1 |
| 820 Ω | 1 | 0.033 μF | 3 |
| 1 kΩ | 2 | 0.1 μF | 1 |
| 2.2 kΩ | 1 | 0.47 μF | 1 |
| 3.3 kΩ | 3 | 1 μF | 1 |
| 4.7 kΩ | 3 | 100 μF[b] | 1 |
| 10 kΩ | 7 | 1000 μF[b] | 1 |
| 22 kΩ | 2 | | |
| 100 kΩ | 2 | | |
| 330 kΩ | 1 | | |
| 1 MΩ | 1 | | |
| 10 MΩ | 1 | | |
| 1 kΩ[a] | 1 | | |

[a]2 W resistor.

[b]50 V capacitor.

### Diodes, transistors, analog IC's

four Si signal diodes;
one 1 A Si rectifier diode;
two 3.3 V 1 W Zener diodes;
one 5.1 V 1 W Zener diode;
one diode bridge element;
one red light-emitting diode;
three 2N3904 NPN transistors;
three 2N3906 PNP transistors;
two VP0610L MOSFETs;
two VN0610N MOSFETs;
two 2N5485 N-channel JFETs;
one 411 op amp;
one 555 timer;
three 741 op amps;
one 311 comparator.

### Miscellaneous

four alligator clips;
two fat-pin adapter sockets.

## Digital components

| Component | Number required |
| --- | --- |
| 7400 | 1 |
| 7404 | 1 |
| 7432 | 1 |
| 7474 | 4 |
| 7485 | 1 |
| 7486 | 1 |
| 7489 | 1 |
| 7490 | 2 |
| 7493 | 1 |
| 74112 | 1 |
| 74121 | 1 |
| 74138 | 1 |
| 74150 | 1 |
| 74191 | 1 |
| 74373 | 1 |
| DAC0806 | 1 |
| ADC0804 | 1 |
| TIL311 | 2 |

## Suppliers of parts

There are numerous companies selling electronic components and supplies. Most allow customers to purchase small quantities directly over the web. Prices are reasonable and service is excellent. Several (e.g., Digi-Key Corp) even have links to product data sheets as part of their online catalog. Product information, availability, and pricing are easily found through a few quick web searches. We've included a few URLs to help get you started.[1] At the time of going to press, the parts and supplies needed to complete the exercises within this book could be purchased from the companies below. Pricing and availability may vary, so shop around!

RadioShack
http://www.radioshack.com/

Digi-Key Corp
http://www.digikey.com/

Newark Electronics
http://www.newark.com/

Tequipment
http://www.tequipment.net/

Electronix Express
http://www.elexp.com/

Arrow Electronics
http://www.arrow.com/

Jensen
http://www.jensentools.com/

---

[1] The publisher has used its best endeavors to ensure that all URLs referred to in this book are correct and active at the time of going to press. However, the publisher has no responsibility for the websites and can make no guarantee that a site will remain live or that the content is or will remain appropriate.

# Appendix B
## Common abbreviations and circuit symbols

### Order-of-magnitude prefixes

$m = $ milli $= 10^{-3}$

$\mu = $ micro $= 10^{-6}$

$n = $ nano $= 10^{-9}$

$p = $ pico $= 10^{-12}$

$f = $ femto $= 10^{-15}$

$k = $ kilo $= 10^3$ (or kilohm $= 10^3\ \Omega$)

$M = $ mega $= 10^6$ (or megohm $= 10^6\ \Omega$)

$G = $ giga $= 10^9$

$T = $ tera $= 10^{12}$

### Mathematical symbols

$\sim$ of order

$\approx$ approximately equal to

$\equiv$ equals by definition

$\Delta$ change in

$\Rightarrow$ implies

### Electrical terms

$\beta = h_{FE} = $ transistor current gain

$\omega = $ angular frequency

$\Omega = $ ohm

$A = $ ampere

$AC = $ alternating current

$C = $ coulomb

$C = $ capacitance

$dB = $ decibel

$DC = $ direct current

$F = $ farad

$f$ = frequency

$g_m$ = transconductance

H = henry

Hz = hertz

$I$ = current

$L$ = inductance

$P$ = power

$Q$ = quality factor (of a bandpass filter)

$R$ = resistance

V = volt

$V_{CC}$ = most positive voltage in a circuit (positive supply voltage)

$V_{EE}$ = most negative voltage in a circuit (negative supply voltage)

$X$ = reactance

$Z$ = impedance

## Electrical devices

| | |
|---|---|
| ADC | analog to digital converter |
| C | symbol used in schematics for a capacitor |
| CMOS | complementary-MOSFET integrated-circuit family |
| CRT | cathode-ray tube |
| DAC | digital to analog converter |
| ECL | emitter-coupled-logic integrated-circuit family |
| FET | field-effect transistor |
| JFET | junction FET |
| L | symbol used in schematics for an inductor |
| MOSFET | metal-oxide-semiconductor FET |
| op amp | operational amplifier |
| Q | symbol used in schematics for a transistor; can also refer to the latched output of a flip-flop or register |
| R | symbol used in schematics for a resistor |
| SPDT | single-pole-double-throw switch |
| SPST | single-pole-single-throw switch |
| TTL | transistor–transistor-logic integrated-circuit family |

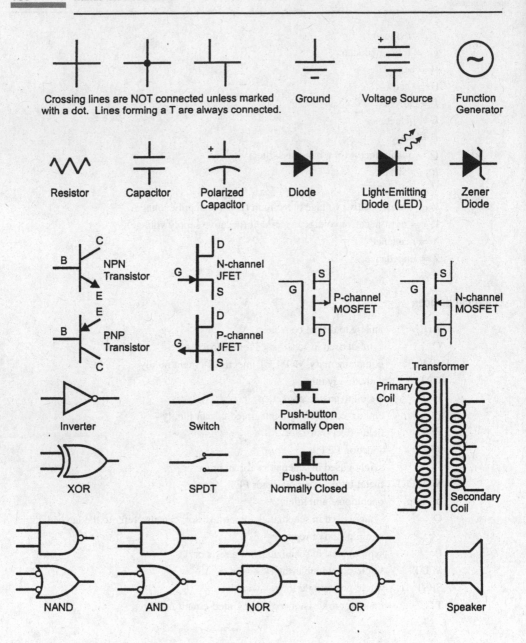

Crossing lines are NOT connected unless marked with a dot. Lines forming a T are always connected.

Ground

Voltage Source

Function Generator

Resistor

Capacitor

Polarized Capacitor

Diode

Light-Emitting Diode (LED)

Zener Diode

NPN Transistor

N-channel JFET

P-channel MOSFET

N-channel MOSFET

PNP Transistor

P-channel JFET

Inverter

Switch

Push-button Normally Open

Transformer

Primary Coil

XOR

SPDT

Push-button Normally Closed

Secondary Coil

NAND

AND

NOR

OR

Speaker

# Appendix C
## *RC* circuits: frequency-domain analysis

In many freshman-physics textbooks, the frequency-domain analysis of *RC* circuits is not explicitly treated; however, it is not particularly difficult. Here is a detailed derivation.

At any moment of time, the charge $Q$ stored on the capacitor is proportional to the voltage $V_C$ across it:

$$Q = CV_C. \tag{C.1}$$

If the voltage across the capacitor is varying sinusoidally in time, it follows that the charge must also vary sinusoidally. Then, since the current $I$ flowing onto one plate of the capacitor is the time derivative of the stored charge, the current must also be a sinusoidal function, but out of phase with the voltage by 90° (since the derivative of the sine is the cosine, which is out of phase with the sine by 90°, and the derivative of the cosine is minus the sine).

Now consider a series *RC* circuit being driven by a sinusoidal AC voltage source (Fig. C.1). Since the resistor and capacitor are in series, they must have the same current flowing through them; however, it is not necessarily in phase with the source voltage, $V$. Suppose (for the sake of definiteness) that

$$V = V_0 \sin \omega t, \tag{C.2}$$

i.e. we have chosen the zero of time to be a moment when the voltage across the source is zero. Then, allowing for an unknown phase difference between the current in the circuit and the voltage applied by the source, we can write

$$I = I_0 \sin (\omega t + \phi). \tag{C.3}$$

Kirchhoff's voltage law tells us that, at any moment of time, the applied voltage must equal the sum of the voltage across the capacitor and that across the resistor:

$$V = V_R + V_C. \tag{C.4}$$

Now, by Ohm's law, $V_R = IR$, and we also have $V_C = Q/C = (1/C) \int I \, dt$. Substituting these relations into Eq. C.4,

$$V = IR + \frac{1}{C} \int_{t_0}^{t} I \, dt \tag{C.5}$$

$$= I_0 R \sin (\omega t + \phi) + \frac{1}{C} \int_{t_0}^{t} I_0 \sin (\omega t + \phi) \, dt, \tag{C.6}$$

where we have made use of Eq. C.3.

We can easily carry out the integration in Eq. C.6 using the substitution $u = \omega t + \phi$, giving

$$V = I_0 R \sin (\omega t + \phi) + \frac{I_0}{\omega C} \int_{\omega t_0 + \phi}^{\omega t + \phi} \sin u \, du \tag{C.7}$$

$$= I_0 R \sin (\omega t + \phi) - \frac{I_0}{\omega C} [\cos (\omega t + \phi) - \cos (\omega t_0 + \phi)]. \tag{C.8}$$

The constant of integration, $\cos (\omega t_0 + \phi)$, can be determined by the condition $V(0) = 0$, which we assumed in writing Eq. C.2:

$$V(0) = 0 = I_0 R \sin \phi - \frac{I_0}{\omega C} [\cos \phi - \cos (\omega t_0 + \phi)], \tag{C.9}$$

giving

$$\cos (\omega t_0 + \phi) = \cos \phi - \omega R C \sin \phi, \tag{C.10}$$

thus

$$V = I_0 R \sin (\omega t + \phi) - \frac{I_0}{\omega C} [\cos (\omega t + \phi) - (\cos \phi - \omega R C \sin \phi)] \tag{C.11}$$

$$= I_0 R [\sin (\omega t + \phi) - \sin \phi] - \frac{I_0}{\omega C} [\cos (\omega t + \phi) - \cos \phi], \tag{C.12}$$

which clearly satisfies $V(0) = 0$.

Eq. C.12 can be simplified using the trigonometric identities for sines and cosines of sums:

$$V = I_0 R [\sin \omega t \cos \phi + \cos \omega t \sin \phi - \sin \phi]$$
$$- \frac{I_0}{\omega C} [\cos \omega t \cos \phi - \sin \omega t \sin \phi - \cos \phi]. \tag{C.13}$$

Gathering and separating terms in $\sin \omega t$ and $\cos \omega t$, and using $V = V_0 \sin \omega t$, since $\sin \omega t$ and $\cos \omega t$ are independent functions of time, we obtain two equations:

$$V_0 \sin \omega t = \left( I_0 R \cos \phi + \frac{I_0}{\omega C} \sin \phi \right) \sin \omega t \tag{C.14}$$

$$I_0 R \sin \phi - \frac{I_0}{\omega C} \cos \phi = \left( I_0 R \sin \phi - \frac{I_0}{\omega C} \cos \phi \right) \cos \omega t. \tag{C.15}$$

Eq. C.15 states that a constant is equal to the same constant times a function of time. This can be satisfied for all times only if the constant is zero,[1] thus

$$\phi = \tan^{-1} \frac{1}{\omega R C}. \tag{C.16}$$

Eq. C.14 can be simplified as

$$V_0 = I_0 R \cos \phi + \frac{I_0}{\omega C} \sin \phi. \tag{C.17}$$

---

[1] Otherwise we could divide through by the constant to obtain $\cos \omega t = 1$, which clearly does not describe the behavior of the circuit.

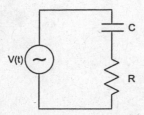

**Fig. C.1.** Series *RC* circuit.

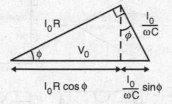

**Fig. C.2.** Right triangle represented by Eq. C.17, illustrating that $V_0 = I_0 R \cos \phi + \frac{I_0}{\omega C} \sin \phi$.

This describes a right triangle with hypoteneuse of length $V_0$ and sides of length $I_0 R$ and $I_0/\omega C$ (Fig. C.2), which is a useful way of visualizing the relationship among the amplitudes of the source voltage, resistor voltage, and capacitor voltage. The relationship is Pythagorean:

$$V_0^2 = (I_0 R)^2 + \left(\frac{I_0}{\omega C}\right)^2.$$ 
(C.18)

We thus have

$$I_0 = \frac{V_0}{\sqrt{R^2 + \left(\frac{1}{\omega C}\right)^2}}.$$ 
(C.19)

If we take the output as the resistor voltage, we get a high-pass filter:

$$V_{\text{out}} = I_0 R = \frac{V_0}{\sqrt{1 + \left(\frac{1}{\omega R C}\right)^2}}.$$ 
(C.20)

If we take the output as the capacitor voltage, we get a low-pass filter:

$$V_{\text{out}} = \frac{I_0}{\omega C} = \frac{V_0}{\sqrt{(\omega R C)^2 + 1}}.$$ 
(C.21)

# Appendix D
## Pinouts

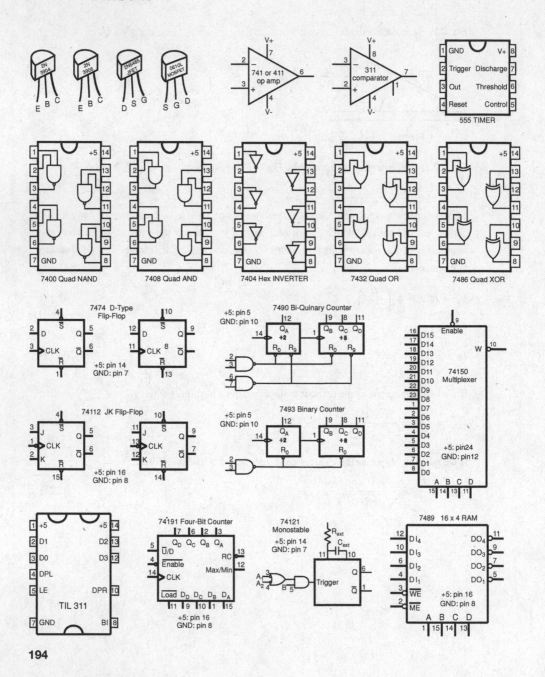

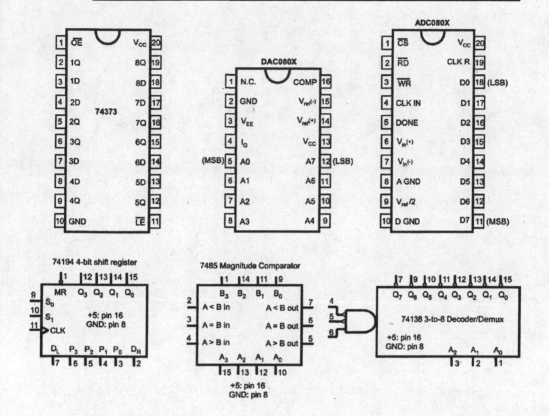

# Glossary of basic electrical and electronic terms

**ampere** Basic unit of current: 1 ampere = 1 A = 1 coulomb/second.

**angular frequency** Rate of change of phase. Measured in radians per second: $\omega = 2\pi f$.

**anode** The negative terminal

**attenuation** Decrease in voltage or current (also implies power reduction).

**capacitor** Device used to store charge and energy. The capacity of a capacitor is called the capacitance. Capacitance $C$, charge $Q$, and voltage $V$ are related by the equation $Q = CV$.

**cathode** The positive terminal

**cathode-ray tube** A large vacuum tube in which the electron beam can be steered to create a visible pattern on a phosphorescent screen.

**charge** A fundamental property of some elementary particles. Electrons have charge $-1e$, and protons and holes have charge $+1e$, where $e = 1.602 \times 10^{-19}$ coulombs.

**common** Voltage reference point (0 V). Also called *ground*.

**compliance** (usually of a current source): Range over which circuit performance is stable.

**coulomb** Unit of charge: 1 coulomb = $6.241 \times 10^{18}\, e$, where $e$ is the charge of the electron.

**current** Rate at which charge flows. Defined as the amount of charge that passes through a given surface (such as the cross-section of a wire) per unit time. A convenient analogy is the rate at which water flows under a bridge or through a pipe. Positive current flows from points of higher voltage to points of lower voltage. (Due to Benjamin Franklin's choice for the definition of positive charge, this is opposite to the flow of electrons.)

**daraf** Unit of inverse capacitance.

**decibel** Unit for specifying a voltage or power ratio on a logarithmic scale.

**dynamic resistance** Effective resistance of a nonlinear element (typically a PN junction), such as a diode or transistor junction.

**farad** Unit of capacitance: 1 farad = 1 F = 1 coulomb/volt. The farad is a large unit; commonly available capacitors range in size from a few picofarads to many thousands of microfarads.

**feedback** A design approach or situation in which an electronic signal communicates information from the output of an electronic device or circuit to its input. Positive feedback enhances a change at the output (i.e., a growing output with positive feedback grows even larger), while negative feedback counteracts a change at the output.

**frequency domain** AC circuit analysis approach that focuses on circuit response to sine waves vs. their frequency.

**gain** Increase in voltage or current (also implies power amplification).

**gate** A circuit that performs digital logic, such as an AND gate or a NOR gate.

**ground** Voltage reference point (0 V). Also called *common*.

**henry** Unit of inductance.

**hertz** Unit of frequency: 1 hertz = 1Hz = 1 cycle/second.

**impedance** Degree to which a circuit element impedes the flow of current; includes both resistive and reactive components. In the standard electrical-engineering notation, resistance is a real quantity and reactance is imaginary, corresponding to their $\pm 90°$ phase difference, thus impedance is given by $\vec{Z} = \vec{R} + i\vec{X}$.

**jack** Connector used to accept a plug; socket.

**Kirchhoff's current law** The net current flowing into or out of any point in a circuit is zero.

**Kirchhoff's voltage law** The total voltage around any closed loop is zero.

**mho** Unit of transconductance; inverse of an ohm.

**ohm** Unit of resistance. 1 ohm = 1 $\Omega$ = 1 volt/ampere.

**plug** Connector that plugs into a socket or jack.

**quiescent** Default voltage and/or current values when an input signal is absent.

**reactance** Capacitive $(X_C)$ and/or inductive $(X_L)$ component of a circuit element's impedance.

**resistance** Degree to which a device impedes the flow of DC current; nonreactive component of impedance. Measured in ohms. (For a nonreactive device, also the degree to which the device impedes the flow of AC current, i.e., for a resistor, $Z = R$.)

**slew rate** Rate at which an output voltage changes.

**socket** Connector used to accept a plug; jack.

**Thévenin equivalent** A way to model complex circuits based on Thévenin's theorem, which reduces most circuits to a single ideal voltage source in series with a single impedance.

**time domain** AC circuit analysis approach that focuses on circuit response to an arbitrary waveform vs. time.

**volt** Unit of electrostatic potential: 1 volt = 1 V = 1 joule/coulomb.

**voltage** Electrostatic potential. Voltage is defined and measured with respect to a common reference or ground point. When multiplied by the value of the charge, voltage gives the potential energy of the charge with respect to that reference. Positive current flows from points of higher voltage to points of lower voltage (from larger potential energy to lower potential energy).

**V$_{CC}$** Most positive voltage in a circuit.

**V$_{EE}$** Most negative voltage in a circuit.

# Index

Printed in the United States
By Bookmasters